P9-AOY-018

Digital Video and Audio Compression

Other McGraw-Hill Books of Interest

Digital Video and Audio Compression

Stephen J. Solari

McGraw-Hill
New York San Francisco Washington, D.C. Auckland Bogotá
Caracas Lisbon London Madrid Mexico City Milan
Montreal New Delhi San Juan Singapore
Sydney Tokyo Toronto

Library of Congress Cataloging-in-Publication Data

Solari, Stephen J.
Digital video and audio compression / Stephen J. Solari.
p. cm.
Includes index.
ISBN 0-07-059538-0
1. Digital video. 2. Video Compression. 3. Sound—Recording and reproducing—Digital Techniques. 4. Coding theory. I. Title.
TK6680.5.S65 1997
621.389′7—dc2 96-30094
CIP

McGraw-Hill

A Division of The ***McGraw-Hill*** *Companies*

1 2 3 4 5 6 7 8 9 0 FGR/FGR 9 0 2 1 0 9 8 7

ISBN 0-07-059538-0

The sponsoring editor for this book was Stephen S. Chapman, the editing supervisor was Paul R. Sobel, and the production supervisor was Claire B. Stanley. It was set in Century Schoolbook by Dina E. John of McGraw-Hill's Professional Book Group composition unit.

Printed and bound by Quebecor Fairfield.

Contents

Preface

What's all the excitement about? Why are phone companies taking equity in entertainment companies? Why are cable operators contemplating massive investments in their inftrastructure? Why are otherwise pragmatic CEOs blathering about the imminent convergence of communications technologies?

The current excitement over digital media may be hard to exaggerate, but it is easy to explain. These activities are motivated by the prospect of cost-effective digital processing for audio and video.

Analog signals are information intensive, requiring high bandwidths and care in transmission. Manipulating signals in the analog domain—creating, editing, or manipulating sights or sound—is possible but cumbersome, thus limiting the number of people who can interact with these media. On the other hand, passive entertainment via analog colog television transmission is available today in 98 percent of U.S. households.

The ever increasing complexity and decreasing costs of integrated circuits are making digital compression cost-effective for a broad number of applications.

Video and audio have joined other kinds of data on computers, creating multimedia PCs. Video can join voice in a range of teleconferencing products. The U.S. has adopted a digital system for high definition television. Broadcasters, cable operators and phone companies are vying to provide new services to the home, including video-on-demand and interactive television.

Only large corporations can afford the capital investments needed to realize some of these systems. However, new technologies have the possibility to change the playing field. Just as PCs have destabilized corporate MIS departments and mainframe manufacturers, and the Internet has provided individuals with broad access to all kinds of information, widespread use of digital media could eventually give individuals greater choices and capabilities.

To help evaluate the consequences of the introductions of these technologies into our life, this book covers the basic technologies, infrastructure, costs, and application sectors for digital audio and video compression.

What's compression? Even the least technical reader opens this book with an idea of what compression is. It is to press something together or to force it into a smaller space—what your trash compactor did to last night's leftovers. The other side of the trick is to decompress—to reconstruct the original with an acceptable level of possible degradation.

The reason to compress and decompress is to better utilize the space or resource available. In the world of electronics, this means to make more efficient use of transmission bandwidth or storage media. Compression can be performed on a signal which is either represented in the analog or digital domain. This book emphasizes techiques which are commonly implemented in the digital domain today, even though some of the principles predate the widespread use of cost-effective digital processing.

Using This Book

Digital Video and Audio Compression can be thought of as the introductory cookbook for those wishing to work with digital media. While conventional cookbooks instruct one on how to deliver dishes to delight the senses of taste and smell, this book discusses the ingredients, processing, and presentation of sights and sounds.

Unlike many of today's popular cookbooks that rely on obscure and esoteric foods and tools, this book covers the basics. While a background in physics or engineering will help readers to understand some of the more technical passages, any interested reader should be able to enjoy *Digital Video and Audio Compression* and come away with an enhanced understanding of terms, tricks, and technology of compression.

Compression is often similar to cuisine in that some preferences are a matter of taste. Compression is always similar to cooking in that it relies on the technique espoused by Terry, the chef of the *Fawlty Towers* television show: "What the eye don't see, the chef gets away wif."

This book is divided into four general sections:

- Chapter 1, *Compression: Why and How*. Chapter 1 describes different ways of electronically representing real-world information that we see and hear. The sections explain why audio and video compression are possible and introduce the principle of *source coding*. The chapter then explores some of the different requirements and characteristics of various candidate applications for audio and video compression.

- Chapters 2 through 7, *Sights*. These chapters first review how humans capture and process visual and motion information, present existing analog broadcast standards, and then explore different techniques for reducing the data associated with visual information. Compression schemes—whether advanced by standards committees (e.g., JPEG, MPEG) or individual companies (e.g., DigiCipher, Indeo)—use different combinations of compression techniques. It is in this respect that the recipe for MPEG-2, for example, can have some of the same ingredients as an alternative proposal.
- Chapter 8, *Sounds*. Chapter 8 first explores how we hear, and then examines sound recording, transmission, reproduction, and audio compression methods.
- Chapters 9 through 11, *Integrated Compression Applications*. Now that we've discussed various schemes and techniques, we can see how different industries are employing combinations of these technologies in a variety of applications and services.

With the pace of technology advancing rapidly, keeping up on technical developments is vital to business success. The goal of this book is to provide readers with a valuable edge in one of the hottest areas today—audio and video compression.

Acknowledgments

Thanks are due to Michael Jacobsen for his contribution to this book, to Rolf-Dieter Gutsman for his contribution to Chap. 4, and to Leo Warmuth for teaching and translating.

Thanks are especially due to Hans Van Weersch for reviewing, editing, and proofreading major parts of the book.

Stephen J. Solari

Digital Video and Audio Compression

Chapter

1

Theory and Practice

1.1 Basic Principles

Our eyes and ears, together with our brain, have an impressive capacity to acquire and process information. Although this ability is great, it is also limited, and most sources of sound, image, and text do not represent the information in the most efficient way. A fundamental concept in economics is the scarcity of resources. The availability of storage media, such as magnetic or optical disk, and transmission channels, such as phone lines or broadcast bandwidth, is limited. Through source coding, it is possible to utilize these valuable resources more efficiently.

1.1.1 Source coding

Source coding may sound esoteric but is in fact commonplace, if not ubiquitous. Any scheme to efficiently represent, transmit, or store information falls under this classification. Examples of source coding include Morse code in telegraphy, the limited bandwidth of speech over telephones, and the sequential projection of still images to create the illusion of motion in a movie theater. Facsimiles use source coding. Compression is a subset of source coding.

1.1.2 Redundancy and irrelevancy

Redundancy and irrelevancy are not only facts of life, they are two fundamental phenomena that make data reduction in source coding possible. Some sources, for example, may deliver signals which are represented with more bits per second than is strictly necessary. In information theory, this is called *redundancy*. Source coding can reduce this redundancy and yield a new representation of the information with fewer bits per second. Because only redundant information has been removed, no "true" information has been lost.

Much information is typically *correlated.* Names in the phone book are listed alphabetically. A sentence that starts in English typically does not end in German. Video is typically correlated in several respects. When one scans along the line of a picture, the colors typically do not change abruptly; hence, they are *spatially correlated.* By contrast, this is generally less true of a page of text, where quick transitions between black and white create letters. Nevertheless, the spatial correlation in a page of text allows effective source coding in facsimile machines.

Similarly, unless there is a scene change, not much changes from one sequential picture in a video broadcast to the next. This is called *temporal correlation.*

1.1.3 Lossless and lossy source coding

In lossless coding, the original message, or signal, can be reconstructed in an identical fashion. The information, thus, can be put through this process a number of times without degradation. This is because in lossless coding, only redundant information is discarded.

Often, however, perfect reconstruction of the input signal is not necessary. In the case of audio and video data, some of the information may be difficult or impossible for a human to hear or see. Such information may be considered irrelevant and can be discarded (lost) during the coding process. This is called lossy source coding. However, the data have then lost some integrity, which introduces distortion to the signal.

1.1.4 Coding and transformation

An example of lossy coding is that a digital "word" is assigned to a corresponding analog value. This can be done in a number of ways, some of which are more efficient than others. We could, for example, assign shorter words to values that occur more often.

It may be desirable to perform a calculation on a section of the signal so that it can be analyzed in another realm, or *domain.* For example, we may map a signal from the temporal to the frequency domain, where compression techniques (transforms) can be applied to the signal, as illustrated in Fig. 1.1.

Transforms are powerful tools for both modeling information content and applying compression principles. In the audio world, a transform lets you look at the content over the audio spectrum. In

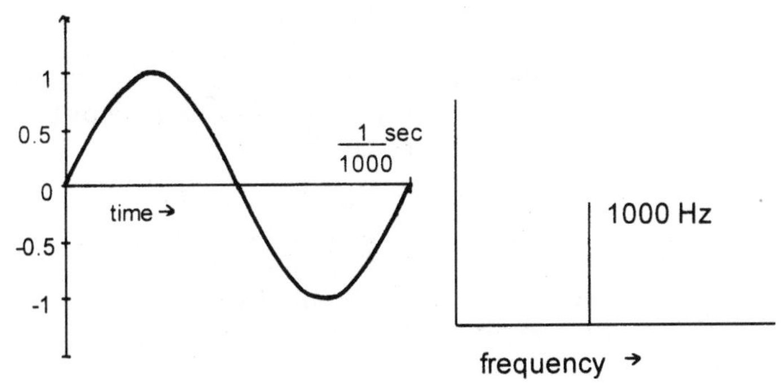

Figure 1.1 A sine wave represented in the temporal and frequency domain.

the video realm, transforms can help to analyze the *spatial frequency* (or extent of detail) in a single picture; it can also be used at cycles per picture height or width. The best known transform for video compression is the *discrete cosine transform* (DCT), which will be discussed in an upcoming chapter. The DCT is a special case of the *Fourier transform.*

1.2 Fourier Transform

Fourier analysis rests on the premise that any periodic waveform can be deconstructed into a number of harmonically related sinusoidal signals of varying amplitudes and phases. Conversely, this same waveform can be reconstructed by properly reassembling these sine waves. This observation allows us to take a waveform that may look quite complicated in the time domain and look at its individual components in the frequency domain.

In the old days (15 years ago) such a transform was often accomplished in the analog domain by sweeping the frequency of a filter applied to a continuous, periodic tone. With the ubiquity of digital signal processing, such work is now performed in the digital domain.

The Fourier transform, like statistics, is a powerful tool. Also like statistics, it can lead to faulty conclusions if not understood and used properly. Remember that audio and video are rarely composed just of continuous periodic tones and that we need to define a window in time or space over which we will apply the transform. Also, the tool is used to evaluate prefilters and postfilters and other signal processors. If it is performed digitally, the assumption is made that the digitization process has added no artifacts.

1.3 *Z Transform*

Another helpful tool, the Z transform, can be derived from a complex spectrum $X(jf)$ multiplied by $\exp(j\times 2\times \pi\times f\times T_a)$ and is equivalent to a time shift of the time-dependent signal by the period T_a. As the delay by T_a is a basic operation, it is appropriate to introduce the equation:

$$z = \exp(j\times 2\times \pi\times f\times T_a)$$

This results in the Z transform, which can be used to clarify the correlation between the output sequence of numbers $y(n)$ and the input sequence of numbers $x(n)$ for the points in time $n\times T_a$. In context, the Z transformation often appears as $z[\exp(-1)]$ or $z[\exp(-n)]$, where $z[\exp(-1)]$ is the signal delay of one clock cycle T_a and $z[\exp(-n)]$ is the delay of n clock cycles.

1.4 Application Requirements

Both audio and video are used in different situations for different purposes. This makes it unlikely that there will ever be one universal, general-purpose compression scheme. Some of the parameters that can vary among applications are discussed below.

1.4.1 Point-to-point versus broadcast

In the case of teleconferencing, two sites are simultaneously transmitting to and receiving from each other. In the case of television, one source can broadcast to millions of receivers. From these two examples, we can see that the ratio of encoders to decoders can vary widely. If there are significantly fewer encoders than decoders, it may be desirable to put more complexity (hence cost) into the encoder.

1.4.2 Real time versus off-line

A news report is an example of a piece of information that needs to be delivered in real time, that is, as it happens. On the other hand, a movie could be compressed over a period of several days as long as it could be decompressed for viewing in continuous time. This information transfer would occur in an off-line mode.

1.4.3 Fidelity

What faithfulness to the original material does the decompressed image need to have? If the image is being used for medical diagnostics, for example, we could reasonably expect a high level of fidelity.

1.4.4 Quality

Quality is a measure of how humans perceive the material. For video, this can include the picture size, resolution, frame rate, and number of possible colors. For sound, quality can include dynamic range, frequency response, and lack of distortion. It has been demonstrated that different levels of quality are acceptable for different applications. The quality of the sights and sounds of a blockbuster film greatly enhances the experience. On the other hand, many are entertained by cartoons such as *The Simpsons,* where writing and stylizing compensate for the low number of colors. Children and adults have spent millions of hours over the last 15 years playing video games with relatively low quality audio and video.

1.4.5 Susceptibility to errors

Some transmission channels or storage media are more reliable than others. When some signals begin to degrade, there is a complete loss of the sound or picture; others degrade gracefully, as analog signals generally do. For example, television receivers that are far away from the transmitter degrade partially, reproducing noisy but understandable audio and video.

1.4.6 Scalability

Is there only one implementation of the application, or do some systems have more decompressing power than others? It may be desirable to create an encoded signal that decompresses very well for people who are willing to pay for high performance but that still delivers some kind of picture to those with less powerful hardware.

1.4.7 Dedicated hardware

We may want to deliver audio or video into an already-installed base, such as a personal computer (PC), without requiring that people go out and buy additional dedicated hardware. In this case, we may select software algorithms suitable to the target hardware platform.

1.4.8 Compression ratio

Considerations of available bandwidth or storage capacity versus desired resolution (spatial and temporal resolution in the case of video, frequency and dynamic range in the case of audio) will lead to a desired compression ratio. While some compression techniques are better suited than others for high compression ratios, it is a safe rule

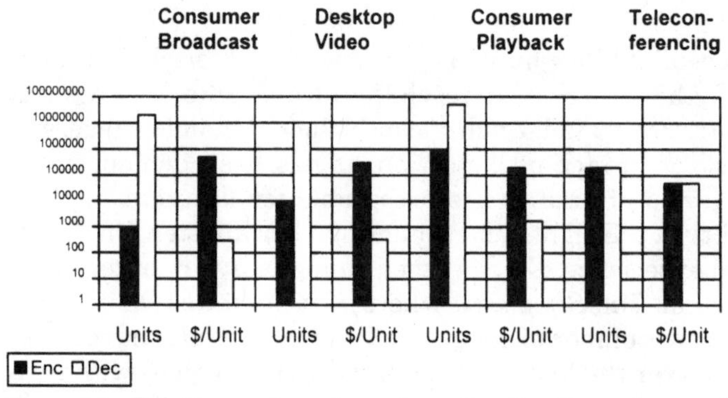

Figure 1.2 Relative costs and number of units of encoders and decoders by application.

of thumb that lower ratios will yield higher fidelity and fewer artifacts.

1.4.9 Cost

Some applications are more sensitive to price than others. While it is easy to imagine that a video production studio would be willing to pay thousands for compression gear, it is generally assumed that it would be difficult to recoup more than $300 in hardware cost for putting a digital video decoder into most households. Figure 1.2 illustrates how encoder and decoder costs are split among popular applications. Four different applications are modeled next, each with different requirements.

- *Consumer broadcast.* Examples of consumer broadcasting include existing terrestrial broadcast, proposed high-definition terrestrial broadcast, and direct broadcast to the home via satellite. In this case, there are hundreds of thousands of receivers for each transmitter. Because the receiver is a consumer item, it will be sensitive to price. For live broadcasts, such as sporting events and news, encoding must be performed in real time.
- *Consumer playback.* Examples of consumer playback systems include various compact-disc–based products, such as those available from Sega, Philips, and 3DO. As consumer products, they are targeted at the $300 price range. Encoding of the signal does not need to be done in real time.
- *Desktop video.* This term can encompass a broad range of configurations. Just as desktop publishing dramatically increased the number of people who could create newsletters and invitations, desktop video is broadening the base of video and multimedia pro-

ducers. Those who are creating compact disc read-only memory (CD-ROM) titles may use audio and video encoding. People who want to enjoy these CD-ROMs will need decoding only. They may or may not have dedicated hardware for the decompression. In 1994, 18 million CD-ROM drives were sold into PC configurations with an average selling price of $1700.

- *Teleconferencing.* Each teleconferencing participant needs both an encoder and a decoder, making teleconferencing a classically symmetrical system. Teleconferencing covers a range of quality, bandwidth, and price points. The connections can range from high bandwidth T1 links through lower bandwidth ISDN connections to the voice-grade bandwidth of conventional analog lines.

1.5 Signals in the Digital Domain

There are two kinds of sources for the sights and sounds which we may want to compress. The first of such sources we find around us in the natural world with our eyes and ears. These sources are normally first converted into an electronic representation in which the signal amplitude is an analog of, for example, the volume or brightness. For digital processing, such as compression, these analog signals must be subsequently converted into the digital domain through sampling and quantizing. This conversion requires some care to avoid any visible or audible disturbances that would alter the information content carried by the analog video or audio signal under consideration.

The second source of signals occurs when data are generated or substantially modified in the digital domain. Examples include synthesized sound and voice, computer graphics, geometrical manipulations and renderings, and even virtual-reality presentations, which may be entirely synthesized. However, these synthesized signals have to obey the same fundamental rules as the previously described digitized analog signals, since their ultimate destination is to be presented to the eyes and ears of a human being.

1.5.1 Synthesized sources

Compression schemes exploit the limited ability of our eyes to see and our ears to hear. Synthesized sounds may have frequencies outside of human hearing range, while computer-generated graphics may have sharp transitions or highly saturated colors that would not survive television transmission and display. For efficient compression, synthesized sources should therefore be limited, prior to compression, to levels that do not exceed the final reproduction vehicle.

Prior to digitization, the signal will have to be constrained to the range of the digitizer and medium, often by limiting peak values or by controlling signal gain. Further issues specific to audio and video preconditioning for compression will be dealt with in their respective sections.

1.5.2 Digitization of analog signals

Analog signals in general are continuous in time and value. The amount of information contained in an analog signal is infinite from the information theory point of view. Clearly, this would make dealing with these signals a tough task in terms of storage capacity and computing power. On the other hand, digitized signals exist only for certain points in time and are represented by discrete amplitude values only. This reduction of information is the key item which makes digital processing so useful and is in fact one of the first steps of compression.

Digitization is a method that reduces the information content to a reasonable level by taking only representative values of the signal under consideration. It does so in two dimensions. It samples in time, and it samples in amplitude. While in theory both steps are independent of each other, in practice they are normally performed by the same processing element, the *analog-to-digital converter* (ADC). It is critical during digitization to capture the useful and desired information contained in the analog signal and remove only the redundancy. Therefore, certain properties of the signals to be digitized have to be known in order to perform the analog-to-digital conversion properly.

1.6 Sampling in Time

Common attributes of video and audio signals include bandwidth, signal-to-noise ratio, signal-to-distortion ratio, and dynamic range. *Bandwidth* describes the possible changes of the analog signals within a given time period, which in turn determines how many samples per unit of time have to be taken to preserve the information content of the signal. The information about the dynamic range and other factors (for example, the noise superimposed onto the signal) determines how accurately the amplitudes of the signals have to be preserved to prevent any noticeable or annoying disturbances.

To convert an analog signal into a digital signal, the analog signal is usually sampled at equal time intervals, and the amplitude of each sample is quantized and assigned to a digital code word. The digital signal is thus a sequence with a constant bit rate derived from the equidistant sampling process of binary-coded numbers of equal length.

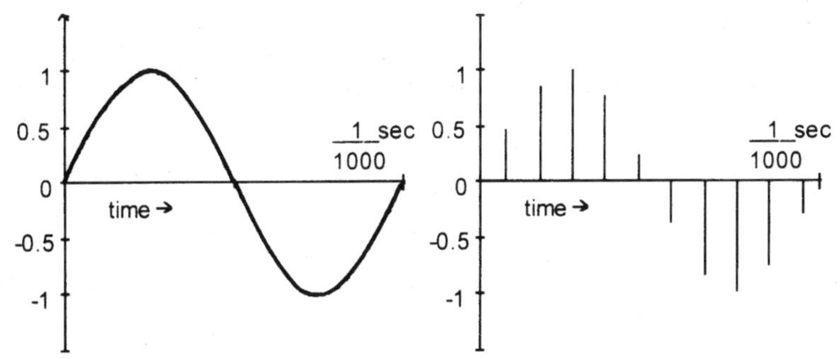

Figure 1.3 A 1-kHz signal represented in the analog and digital domain, showing temporal quantization.

If the clock used to measure the time intervals is unstable, samples will not always be taken at equal intervals, which will distort the signal. If this jitter is random, it can contribute a gaussianlike noise. However, the clock may in fact have another signal [such as a 60-hertz (Hz) power supply] mixed in with it that causes a periodic cycle in the clock.

Figure 1.3 demonstrates the principle of digital signal processing. The analog time-dependent input signal $x(t)$ passes through the antialiasing filter. This is followed by a sampler, which is a sample-and-hold circuit within the ADC clocked at the sampling rate with the frequency f_s. The sampler converts the analog signal into a time-discrete signal, which then—in the following section of the ADC—is quantized and represented by a binary code. The overall process is illustrated in Fig. 1.4.

The Fourier transform of the analog input signal $x(t)$ after sampling shows that its complex spectrum $X(f)$ is periodic: the spectrum is repeated with the interval of the sampling frequency f_s. If the spectrum of the input signal is not at least half of the sampling frequency, which would be the case if it is not sufficiently band-limited or if f_s is too small, frequency bands will overlap. This effect is called *aliasing* (F_A) and occurs in the hatched areas shown in Fig. 1.5. Because of the

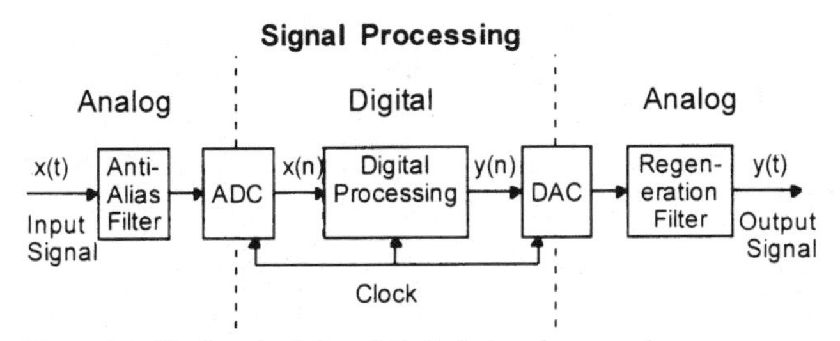

Figure 1.4 Basic principles of digital signal processing.

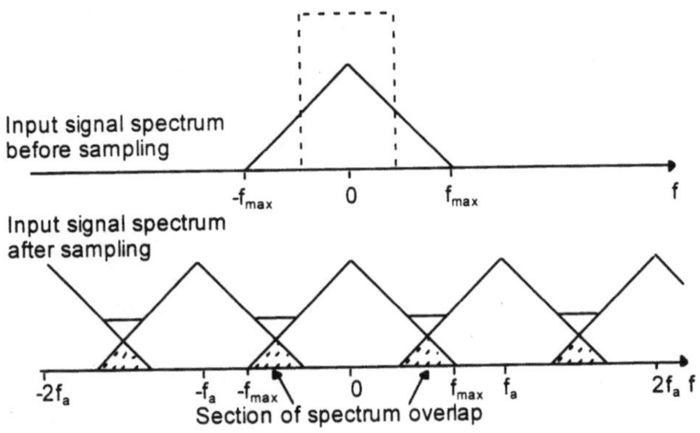

Figure 1.5 The effect of sampling on the input signal spectrum without an antialiasing filter.

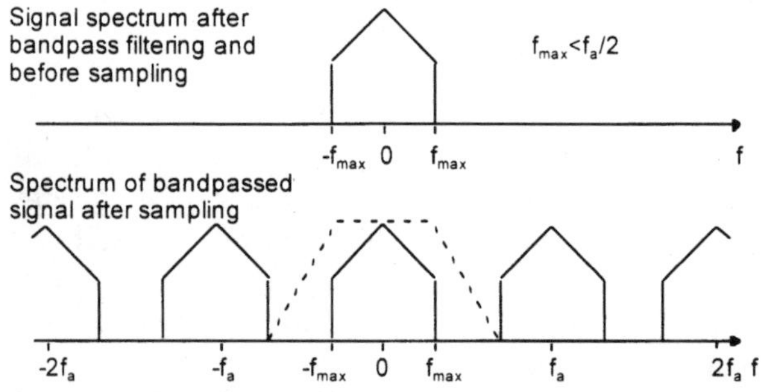

Figure 1.6 The effect of sampling on the input signal spectrum with a proper antialiasing filter.

spectral overlap shown, the original analog signal cannot be regenerated correctly from the sampled signal because the spectral content belonging to the original signal and that of the sampled signal can no longer be distinguished from one another.

Figure 1.5 shows the spectrum of an analog input signal before and after sampling if the sampling theorem is not satisfied: the cutoff frequency $f_{\max}$ of the input signal is greater than half of the sample frequency. It is only by using an antialiasing filter that limits the spectrum to a new $f_{\max}$ that the sampling theorem is satisfied. This is shown in Fig. 1.6; there is no overlapping in the spectrum because the bandwidth of the input signal has been limited to

$$f_{\max} < \frac{f_a}{2}$$

It is therefore possible to regenerate the analog input signal $x(t)$ from the spectrum between $+f_{max}$ and $-f_{max}$ of the sampled signal by appropriate filtering, even though the signals between the sampling points were lost during sampling. If this condition is satisfied, no signal corruption or information loss results from the sampling procedure. This sampling theorem is usually attributed to C. E. Shannon, who stated it in his information theory during the 1940s.

In cases where the input signal is not already band-limited sufficiently, the low-pass antialiasing filter is inserted before the ADC to limit the input signal frequencies to less than $f_a/2$ for a given sampling rate f_a. It is important to note that the limitation not only has to be applied to the relevant part of the signal—with respect to the information content—but also to any superimposed noise and any other disturbing signals occurring at higher frequencies. Such signals may lead to the same kind of unwanted and irreversible alias effects that would result from failing to limit the bandwidth of the original signal. Filters with perfect cutoff characteristics, however, are not realizable in practice, so f_a must always be chosen considerably higher than double the highest useful frequency f_{max}, where f_{max} determines the resolution of the picture or the bandwidth of the audio signal.

When f_a is much greater than two times f_{max}, it is easier to control the behavior and to design the filters because their rejection of higher frequencies does not need to be so steep. Care must also be taken that the sampler and the ADC do not themselves generate excessive high-frequency noise, which could occur, for example, if the clocking signal is mixed with the analog signal.

It can be assumed that the bandwidth of the signal will not be extended during signal processing. It does not matter if the bandwidth of the signal is widened a little. Small changes in the bandwidth occur during additions and subtractions of data signals or constants. Multiplication of signals may be necessary, however, and this can cause considerable widening of the output bandwidth. The correct choice of sampling frequency and sufficient limiting of the bandwidth will prevent overlapping of the subspectrums.

Sometimes it is necessary to alter the sampling rate in the digital domain. Samples are then needed at new instants that are between known sampling points. These signals can be calculated using an appropriate interpolation algorithm which has the structure of a *finite-impulse-response phase-linear* digital filter.

Before summation within the filter, the initial values of the samples $x(k \times T_a)$ are multiplied by the interpolation coefficients for all k. These interpolated coefficients are the values of delayed weighting factors C_k

$$C_k = si\left(\pi \times \frac{t - k \times T_a}{T_a}\right)$$

with the sampling period $T_a = 1/f_a$. Thus, the *si function* is defined by

$$si\ (x) = \frac{\sin x}{x}$$

which is the inverse Fourier transform of an ideal rectangular low-pass filter characteristic in the frequency domain. It is also called the *Whittaker reconstruction.* Because the weighting factors decrease quickly for high values of k, it is sufficient to make the approximation with a limited number of the primary samples.

1.6.1 Sampling amplitude

The signals are still analog after being sampled in time and must be quantized and coded appropriately before they can be processed by the digital signal processor. Quantization and encoding are performed within the ADC.

At the output of an ADC, the instantaneous value of the analog signal sample is described by a binary-coded value of finite and constant length. Conversion to discrete amplitude levels can introduce the following errors:

- *Rounding errors.* These can cause deviations from the ideal behavior of the system.
- *Quantization noise* (*rounding noise*). This affects the output signal and can create an instability (*limit cycles*) limitation by going off-scale.

A compromise must be found between the quality of processing and the complexity and price of implementing a system with extended word lengths. The bit resolution can be increased by interpolation at the cost of time resolution. By simple averaging of $2n$ successive samples, under certain conditions statistical methods say that word lengths will be extended by n bits. However, averaging will cause a loss of high-frequency information if special precautions are not taken and should therefore be avoided. Figure 1.7 illustrates the quantization process.

As a criterion for visibility and audibility of quantization effects, simple statistical methods are often inadequate because they do not cover special cases. For example, very small changes in the gray level of video signals during quantizing may easily lead to annoying contouring effects, depending on the subject matter.

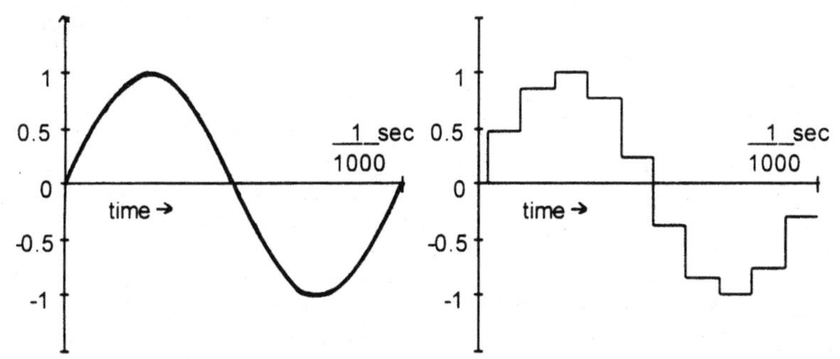

Figure 1.7 A 1-kHz signal represented in the analog and digital domain, showing amplitude quantization.

1.6.2 Pulse-code modulation

Signals that have been sampled and quantized as described previously are called *pulse code modulation* (PCM) signals because each sample is coded independently of any other sample with code words of constant length. Each code word usually consists of several bits, with 8 to 10 bits typically used for video signals. While 8 bits may be used for low-grade audio, 16 to 20 bits are needed for high-fidelity sound.

Transportation of the code words between processing blocks may be done for all bits in parallel at the sampling rate f_a, so that each block has access to all bits of a code word at the same time (which is advantageous for processing). Or, the interface may be handled as a serial bit stream to save on cabling and connection points. The latter method, however, requires another clock frequency, which is a multiple of the sampling rate times the bits per code word (for example, $16 \times f_a$ for 16 bits) and time division multiplexing at the encoder and corresponding demultiplexing at the decoder. Because putting the bits in series reduces the number of physical connections needed while having no influence on the information contained in the code words, serialization is often used for digital transmission.

1.6.3 Digital processing

The ADC output signal is delivered to the input of the digital processor. The digital signal processor operates on the digital numbers $x(n)$ according to a system algorithm, and outputs the resulting series $y(n)$ (or several series of numbers) at the sampling rate. The system algorithm determines the system response and is chosen according to the required operating characteristics.

In case of subsequent digital-to-analog conversion and rendering, the principles of a digital signal processor are similar to those of a real-time digital computer. It is important that the digital signal processor operates in real time because uninterrupted signal processing has to be guaranteed. With every clock cycle, an operation is performed according to the system algorithm. Because the throughput time often is very short, the internal signal processing of the data words usually is performed in parallel.

In the case of synthesized signals, the synthesis also has to be done according to the previously stated rules; the new information either has to be prepared and stored in a buffer memory for real-time readout or must be created in real time where—especially for video signals—a considerable amount of processing power is necessary to meet the real-time requirements.

The digital processors, whether programmable or not, provide two basic structures for filtering, manipulations, and related tasks.

- *Nonrecursive systems.* These systems have no feedback. Included in this class are *finite-impulse-response* (FIR) filters, which are stable and can easily provide linear phase response.
- *Recursive systems.* These systems contain feedback loops. For *infinite-impulse-response* (IIR) filters, stability is a critical issue during design because the pulse response does not disappear after a finite number of clock cycles.

The combination of both systems, together with intelligent use of large memories and lookup tables for nonlinear operations, allows a large variety of operations to be implemented.

For all operations, especially for the synthesis of sound and pictures, the properties of the digital system have to be taken into account. The sampling theorem must be strictly observed, and violations must be rigorously avoided: all operations must be performed in a way such that frequencies higher than given by the limit f_g, or at least by half of the sampling frequency ($f_a/2$), cannot occur.

Once higher spectral components are generated, they are immediately folded down in the frequency domain by mirroring at the sampling rate f_a, becoming part of the repetitive spectrum of the sampled signal. Once created, they cannot be removed, because there is no chance to distinguish between wanted and these kind of unwanted signal components in the digital domain.

This leads to the requirement given previously: both former analog signals and synthesized signals must obey the same rules with respect to the frequency domain. This underlines the importance of always considering the spectrum constraints when working in a digital system.

1.6.4 Preparation of analog rendering

The digital output from the processor has to be transferred back into the analog world of the video screen and the speakers. The digital code words must be converted back into analog signal values by the *digital-to-analog converter* (DAC), which delivers one value from the range of possible output levels at each clock cycle, giving a continuous signal in time. The only unwanted effect is that with amplitude changes between successive samples there is no smooth transition between the two amplitude values but rather a step, which may lead to distortions. Even if the sampling frequency and its multiples are suppressed by the *hold* function during one sampling period of the DAC, the repeated spectra of the signal carrying the information are still present (the degree of suppression follows a si function with zeros at f_a and its multiples). Therefore, the analog output signal from the DAC must be passed through a *regeneration filter.* This is a low-pass filter which, if designed properly, has no influence on the information carried by the signal up to the critical frequency. It functions in a manner analogous to the antialiasing filter at the input by suppressing high-frequency signal content above f_g, which is still present from the sampling. The requirements for the regeneration filter are similar to those of the antialiasing filter, but if the circuitry after the DAC has sufficient attenuation above the required cutoff frequency, the regeneration filter sometimes may not be necessary as a separate component. Some of the postfiltering can even be done in the digital domain: *oversampling*—a well-known technique in digital audio—can be applied in the digital domain by using a much higher sampling rate only at the back end of this process, moving artifacts out to multiples of the sample clock and reducing the requirements for analog postfiltering.

1.7 The Human Visual System

What you see is what you get—the final destination of a video signal is the human visual system. An understanding of the characteristics of the destination is needed in order to design a data-transmission system using compression—after all, the only information which must be carefully transmitted is that which is actually perceivable by the viewer. Transmission of more information than this increases the data rate with no benefit to the viewer. On the other hand, the transmission of too little information will be noticed by the viewer and regarded as disturbing. Because picture quality is essentially judged by the viewer, the acceptance or rejection depends on the consideration given to the visual perception characteristics during encoder and decoder design.

If the perception characteristics of the human visual system are ignored, the following could occur when designing the transmission system:

- Because the data rate of moving pictures often found in television can be much larger than the assigned channel capacity, greater raw compression factors would have to be achieved than if the limitations of human vision had been considered. Unfortunately, a much higher compression factor cannot be achieved without a reduction in picture quality in the form of coding errors.
- Visible coding errors may occur.

For example, a possible application of picture compression is the digital transmission of TV pictures with good quality on a limited bandwidth terrestrial TV channel. We can estimate the required compression in the following fashion:

$$\frac{\text{Source data rate}}{\text{channel data rate}} = \text{compression factor}$$

$$\frac{160 \text{ Mbit/s}}{20 \text{ Mbit/s}} = 8 \text{ times}$$

Despite intensive effort, no compression technique has yet been found which can guarantee a compression factor of greater than 2 for certain TV pictures or sequences without some loss. If the compression factor is increased even further, all known techniques will generate changes in the picture. Unfortunately, these changes may not be limited to an acceptable loss of irrelevant information that the eye does not perceive.

1.7.1 Picture quality

If there were a universal way with which the quality of a transmission technique could be measured and given as a single parameter, one could directly compare all coding methods. One way of approaching the problem is to measure the *mean square error* (MSE) in the decoded gray level compared to that of the original gray level. Whether or not this can be used as a criterion will be tested by an experiment.

In Fig. 1.8, the test picture is quantized with a resolution of only 6 bits; therefore, only 64 possible gray levels exist. The mean error is approximately 1 gray level/pixel, relative to the original 256 gray levels which are supposed to give very good picture quality. Nevertheless,

Figure 1.8 Picture quantized with 64 possible gray levels, giving a large MSE.

there is no detectable loss in the quality of the reproduced (printed) picture. In Fig. 1.9 the picture is quantized with 8-bit resolution, but during transmission an error has crept in which is nearly unmeasurable by looking at the MSE of the whole picture. However, the picture has probably lost some of its appeal! The total human visual system reacts very sensitively to changes in key information, such as the absent incisor in this example.

What may help here is that the critical points in the head and shoulder areas of pictures for video conferencing are recognized by advanced techniques which therefore locally produce a transmission with better quality at the cost of a locally higher bit rate.[4]

On the other hand, there are coding errors which we are now used to. The interlace used in conventional television transmission is the attempt to compress data using analog techniques, although such phenomena as line flicker and line crawling occur as a result.

From this experiment, we can deduce that during picture coding, the MSE can, with reasonable interpretation, be a useful help, although this criterion alone is not definitive. Primarily, the viewer observes and judges the picture locally and the more artificial the effect of coding errors, the easier they are to detect.

Certain compression techniques usually generate specific and individual coding errors. A rough quantization leads to noise with certain

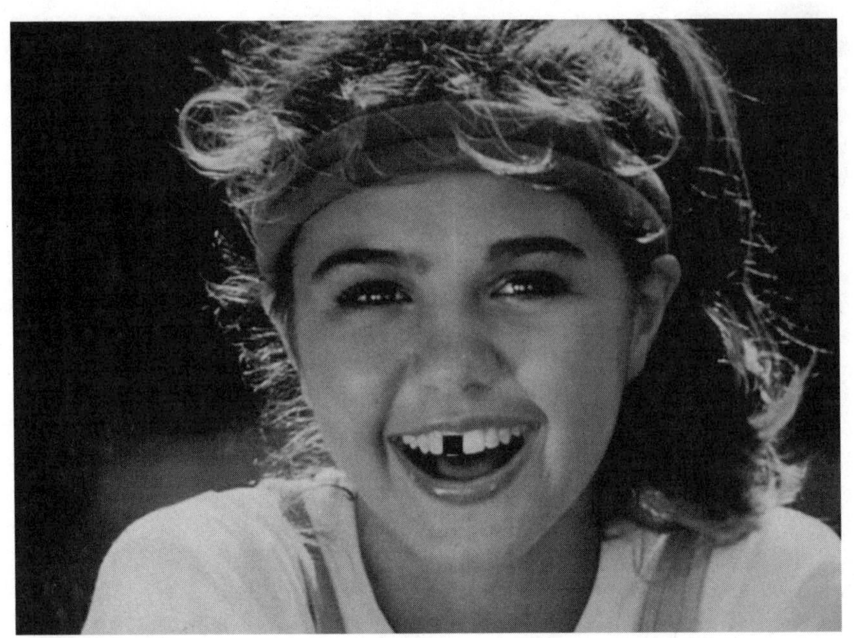

Figure 1.9 Picture quantized with 256 possible gray levels and with visible artifacts, giving a small MSE. (Note: this is a simulated error intended to illustrate a worst-case scenario.)

techniques, which under certain conditions can cause large area structures to be formed in the image, whereas with other methods the data rate is reduced at the cost of picture sharpness. With concepts which work linewise, picture lines lying above each other are compressed consecutively and separately. Through this separate handling, inconsistency in the vertical direction can occur such that the vertical edges of the object in the picture become distorted. Contrary to this, other coders work in a block-oriented way. Through the independent encoding of small blocks of the picture (frequently 8×8 or 16×16 pixels large), distortion can occur due to the visibility of plate-like structures at the block boundaries, known as *blocking effects.*

Other coding errors, such as edge distortion, dot crawling, or plate-like structures, are admissible when they fall below the perception level. And that is the entry point: the search for thresholds of human perceptions below which coding errors will not be detected.

1.7.2 Modulation transfer function

Everyone can remember the eye-test experiment: the larger letters of the test card are immediately recognizable, the smaller ones require some time to ponder, while the smallest appear to be blurred dots only. In a similar way, we sort out at first the larger structures, which, like the large letters, change only slowly per unit of the viewing angle—the information is contained in the low spatial frequencies. In the next step, the finer and, with respect to the viewing angle, the more rapidly changing letters are recognized. Then, even the smallest objects consisting of the highest spatial frequencies will be recognized. And finally, even the most eagle-eyed among us is unable to resolve structures above the spatial frequency, where it is no longer necessary to transmit information. This spatial frequency can be regarded as the frequency limit of the modulation transmission function of the eye. But the eye-test experiment tells us much more. The nearer one approaches the frequency limit, the less one will recognizes not only the useful information but also the coding errors. Coding techniques can be exploited so that picture information corresponding to higher frequencies is quantized more coarsely. Although it has been shown that the eye works in a nonlinear way (it cannot perceive and evaluate individual spatial frequencies of the picture signal), the insensitivity of the eye toward higher frequencies through coarser quantization of the spatial frequencies can be exploited and as a result a lower bit rate can be employed.

1.7.3 The oblique effect: Horizontal and vertical perception

During their development, our senses—including our sight—have adjusted to the special requirements of the human environment. As our surroundings for the most part comprise horizontal and vertical structures (take a brief look around you), the signal processing in our brain is optimized toward this. Because of the lack of stimuli, there is a shortcoming in the recognition of diagonal structures. This so called *oblique effect* is more readily traceable to pattern recognition mechanisms than to a reduced modulation transfer function in a diagonal direction. But even this effect is exploited; portions of the picture with diagonal spatial frequencies which correspond to structures in a diagonal direction can be quantized even more coarsely than the modulation transfer function might otherwise allow.

1.7.4 The Mach effect and contour perception

It is not the surface areas of pictures that hold information but the edges which help us to identify objects. The important part of an object—its edge—is therefore of prime importance in how it is perceived. For this reason, the human visual system recognizes an edge enhancement and steepening; a jump in the brightness of one edge

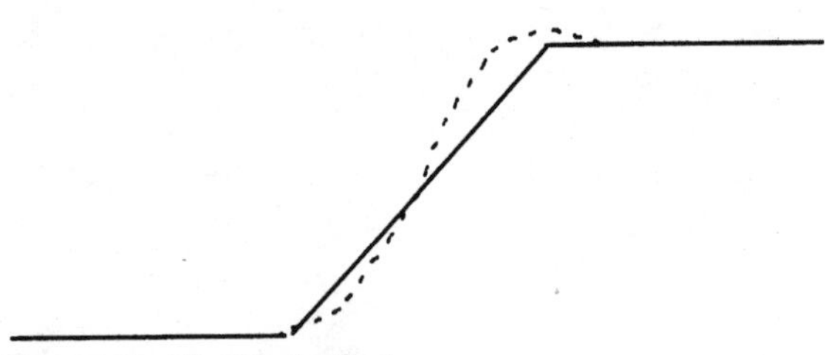

Figure 1.10 The Mach effect.

Figure 1.11 Test picture for the Mach effect.

will result in it being exaggerated, as shown in Fig. 1.10. This is called the Mach effect. Because we have grown used to seeing such edge types, this effect is not noticeable in everyday life. But if this effect is deliberately sought , it will certainly be found. If you casually observe the change between the light and dark areas in Fig. 1.11, you will notice this effect. A thin strip appears on the light side near the change, which is slightly lighter than the paper on the right of it, and slightly darker on the right. Why is this effect important for picture coding?

Edge steepening is a signal-faking process. We can make use of small errors in brightness steps for the representation of edges without them being noticed. It is permissible, for instance, to save bits by transmitting the brightness gradation of an edge with only a few gray levels. One method worth looking at which makes use of this effect, is the *adaptive dynamic range coding* (ADRC) method.[5] The picture is divided into blocks, and first the lowest and the highest gray level of each block is transmitted. Sixteen gray levels are then allocated for the gray value of each pixel within these limits. This process, therefore, gives 4 bits per pixel. Whereas the gray level only changes slowly in the surface areas of pictures, requiring only a few different gray

Figure 1.12 Example of where coding artifacts can be hidden. Area *a* is the small box at top left; area *b* is the small box below and to the right of area *a*.

levels for error-free transmission, this is not the case for edges. But although the edges comprise a series of large steps for the observer, due to the Mach effect, the error is seldom noticed.

1.7.5 Masking

The phrase "looking for a needle in a haystack" demonstrates clearly that irregular structures can completely hinder the perception of embedded, but more or less hidden objects, contained within them. As an example, consider Fig. 1.12. In area *a,* where there is very low picture activity (blue sky), coding errors are immediately recognizable, whereas in area *b,* where there is high, irregular activity (a flowered pasture) errors are disguised by the masking effect.

To elucidate, noise has been added to Fig. 1.13. The noise is clearly detectable in the face of the girl, but not in her hair. This shows that we can disguise coding errors with one mechanism showing the areas with a high level of irregular picture activity—also called texture—and thus save bits. However, a word of caution: coding errors can show up as regular structures (blocking) such that a considerably lower perception threshold can occur. That the block structures can be recognized as an integral grid depends on the pattern recognition capability of the human visual system (Fig. 1.14). When the frame

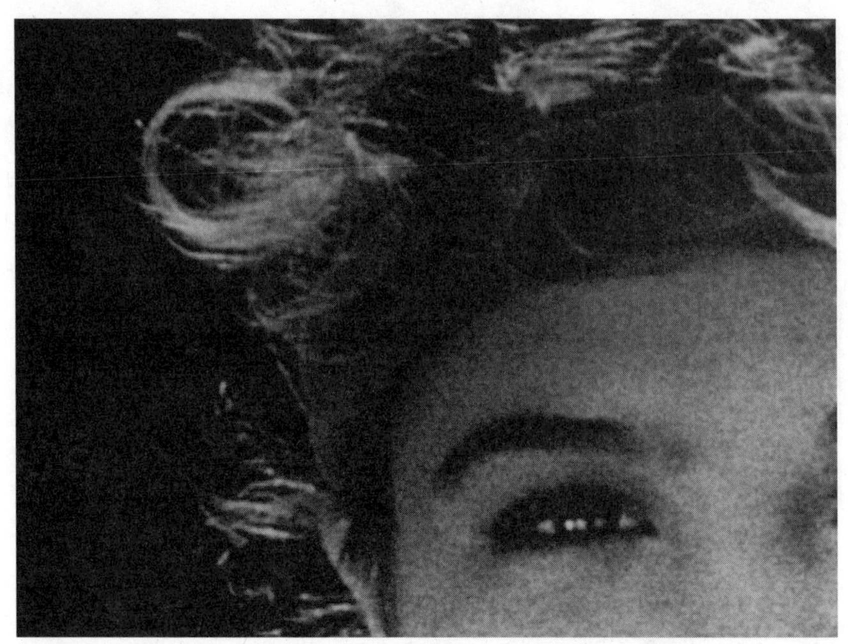

Figure 1.13 An example of structured noise hidden in the woman's hair.

Figure 1.14 Possible artifacts from a block-oriented coder.

Figure 1.15 Column with break (left) and tilt (right).

content moves as a result of motion of the camera, the block structures stay in the same position, but the moving scene and the stationary block structures are separated in the brain and we view the scene through some kind of small windows generated by the coding error. In fact we speak of a "dirty-window effect;" through the relatively independent perceptions of coding errors and scene, the masking effect is no longer effective, even in the area of high picture activity.

Depending upon the image, the eye may react sensitively to changes in the position of an edge. As illustrated in Fig. 1.15, one of the represented blocks happens to have a defect in the middle, which is noticeable even with the smallest displacement. To show another effect, a block from the middle has turned in relation to another. The defect is immediately noticeable despite the fact that the block has only turned by 2°.

A practical example will further illustrate this point. As a first step, some coding methods use parts of the previous picture in time to create the following picture in that they cut the previous picture into squares and arrange them until they look, as much as possible, like the previous picture. Because the previous picture is familiar to the viewer, only rearranged information must be transmitted in order to adequately describe the picture that follows it and so relatively few bits are needed for transmission. In Fig. 1.16 an attempt is made to show this with the scene of a turning spoked-wheel. Because the rotation is only being represented in a poor way by means of displacement, the spokes of the wheel give a torn effect.

(*a*)

(*b*)

Figure 1.16 Torn edges resulting from coding: (*a*) large errors, (*b*) small errors.

Figure 1.17 With a narrow (4:3) aspect ratio, the eye cannot follow large moving objects long enough to detect errors.

1.7.6 Loss in resolution with moving objects

Up to now, the vigilance of the characteristics of the eye with respect to position have applied. Now the characteristics in time will be shown. Imagine that you are standing at the entrance to an arch and are viewing the street so that only a small part of the street scene can be perceived. Such a limited field of view is offered by the traditional television system, as shown in Fig. 1.17. You only see the cars rush past you and you do not have a chance to see the license plate or recognize the driver. For the frame coding, this means that fast-moving objects with large coding errors (with more limited horizontal and vertical resolution) can be transmitted with less bits.

This applies to standard television. With wide-screen high-definition television (HDTV), you in effect step out of the arch and onto the street, acquiring a wider field of vision, as illustrated in Fig. 1.18. The cars no longer rush by; now we have the time to seek an interesting object and to follow it. At the moment the image of the car breaks onto the screen we can identify all the auto's details. Viewers would find it disturbing if, unexpectedly, a red Ferrari appears as a spongy red patch. Because the transmitter does not know in advance which object the viewer will choose to observe, it only makes sense to use this effect for objects with high speeds. We can see clearly by means of

Figure 1.18 With a wider aspect ratio (16:9), the eye can follow a large moving object.

this example that coding for different applications (in this case standard TV in comparison to HDTV) must be treated differently.

1.7.7 Interpolation of sampled movement

When we view television, our senses convert a succession of individual phases of motion into a continuous movement. To save bits, we can transmit slow or stationary objects only relatively seldom but with a high spatial resolution. They are then shown repeatedly and successively at the receiver. Perception of continuous motion, however, ceases with increasing speed of motion. Below a certain level of frames per second the fast motion will appear jerky. This effect can be seen in cartoon films where frames have been left out to save money, or when cinema projectors show films at a standard 24 frames per second. Interframe interpolation can be carried out at the receiver to reduce the jerky motion. In the most simple case, this comprises estimating the intermediate phase by averaging. By this method, however, additional disturbances such as unsharp pictures can occur. Therefore, if we want to avoid jerky motion, we must perform perception tests to carefully determine the switching parameter for the number of necessary motion phases dependent on object speed, size, and contrast.

1.7.8 Moving and stationary picture transmission coding

A system which is optimized for the transmission of stationary pictures can deliver unusable results for moving pictures and vice versa. A stationary picture coder would be optimized for maximum sharpness because the eye has time to view the stationary scene. An observer would perhaps not even notice certain errors, such as a slight displacement of object edge or platelike structures in irregularly structured areas of the picture. Consequently, this artifact of the compression algorithm is perfectly allowable. If this coder is used for the transmission of a moving picture, the result will be intolerable disturbances due to the object edges jumping backward and forward and thereby producing flicker, or the block structure (dirty-window effect) will suddenly appear. Conversely, the moving picture coder would not work effectively as a stationary picture coder because, for example, of the requirement for suppression of block-oriented coding errors within structured picture areas.

The characteristics of our senses with respect to time are more complex than can be covered within this framework, and so the interested reader is referred to further literature on the subject (see Refs. 1 through 3). To aid in the understanding of information presented in following chapters, a list of common abbreviations is given in Table 1.1.

TABLE 1.1 Common Abbreviations

ADM	Adaptive delta modulation
ADPCM	Adaptive differential pulse-code modulation
ADSL	Asymmetrical digital subscriber line
ALC	Automatic level control
AM	Amplitude modulation
ASPEC	Adaptive spectral perceptual entropy coding
ATM	Asynchronous transfer mode
CATV	Cable television
CBP	Coded block pattern
CCIR	Comite Consultatif Internationale des Radiocommunications (International Radio Consultative Committee)
CCITT	Comite Consultatif Internationale de Telegraphique et Telephonique (International Telegraph and Telephone Consultative Committee)
CD	Compact disc
CDI	Compact disc interactive
CD-ROM	Compact disc read-only memory
CD-ROM-XA	Compact disc read-only memory extended architecture
CIF	Common source intermediate format
CRT	Cathode ray tube
CVBS	Composite video, blanking, and sync

TABLE 1.1 Common Abbreviations (*Cont.*)

D2	Color-burst locked digital video format
DAT	Digital audio tape
DAV	Digital audio video (bus)
dB	Decibel
DCC	Digital compact cassette
DCT	Discrete cosine transform
DLC	Digital loop carrier
DSP	Digital signal processing
DVD	Digital video disc
ECS	Entropy-coded segment
EISA	Extended industry standard architecture
EOB	End of block
FM	Frequency modulation
FTTC	Fiber to the curb
GOB	Group of blocks
HDCD	High-density compact disc
HFC	Hybrid fiber/coax
HSI	Hue, saturation, and intensity
ISA	Industry standard architecture (bus)
ISO	International Standards Organization
JPEG	Joint Photographic Experts Group
MD	Minidisc
MPEG	Moving Pictures Experts Group
MUSICAM	Masking-pattern-adapted universal subband integrated coding and multiplexing
NICAM	Near-instantaneously companded audio multiplex
NIU	Network interface unit
NTSC	National Television Standards Committee
PAL	Phase alternating line
PASC	Precision adaptive subband coding
POTS	Plain old telephone service
QAM	Quadrature amplitude modulation
QSIF	Quarter source intermediate format
RGB	Red, green, blue
SBC	Subband coding
SDSL	Symmetric digital subscriber line
SIF	Source intermediate format
SNR	Signal-to-noise ratio
SPL	Sound pressure level
SVHS	Super VHS
VAFC	VESA advanced feature connector
VESA	Video Electronics Standards Organization
VHS	Video Home System (trademark of JVC)
VL	VESA local (bus)
VLC	variable-length coding
VMC	VESA media channel
VOD	Video on demand
YUV	Luminance, chrominance

1.8 References

1. Korn, A., *Picture Processing via the Visual System,* Springer, Berlin Heidelberg, New York, 1982.
2. Wendland, B., *Basic Television Techniques,* Huethig, Heidelberg, 1988.
3. Netravali, A., and B. Haskell, *Digital Pictures Representation and Compression,* Plenum Press, New York and London, 1988.
4. Badique, E., *Knowledge-based Facial Area Recognition and Improved Coding in a CCITT-Compatible Low-Bitrate Video-Codec,* Picture Coding Symposium, Cambridge, March 1990.
5. Kondo, T., et al., *Adaptive Dynamic Range Coding Scheme for Future Consumer Digital VTR,* IERE Proc. 7th Int. Conference on Video, Audio and Data Recording, York (UK), March 1988, pp. 219–226.

Chapter

2

Processing Video for Compression

2.1 Scanning

Before a picture can be transmitted it is necessary to scan the image and to convert the brightness of a specific picture element into an electrical representation, for example, by means of a photo-sensitive cell. Scanning (under existing broadcast standards) is done line by line from the left to the right and from the top to the bottom, just as we read a sheet of paper. The resulting electrical signal can be used to modulate a high-frequency carrier for subsequent terrestrial broadcast or cable transmission, or it may be directly displayed or recorded.

2.1.1 Synchronization

For the picture to be correctly reconstructed on a cathode-ray tube, the reproducing electron beam of the tube must be synchronized with the scanning beam of the photo-sensitive element. This ensures that the transmitted brightness is positioned at the same place in the received picture.

The deflection circuits on the receiver side, therefore, must be synchronized; to ensure this correspondence, synchronization signals are added to the pure brightness information. A *line synchronization* pulse indicates the end of a line, and a *field synchronization* pulse indicates that the bottom line of a displayed picture has been given. To avoid these auxiliary signals producing any brightness on the screen, they are carried by an electrical voltage below normal black, i.e., *ultrablack* (Fig. 2.1.) It is important for the receiver that the detection of line synchronization pulses is not inhibited during the field synchronization pulse.

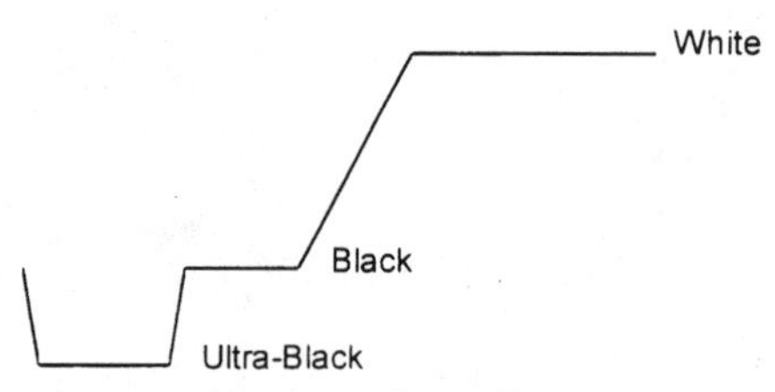

Figure 2.1 A horizontal scan line.

2.1.2 Interlace

In present broadcast schemes, the lines of a complete picture (frame) are not scanned and displayed line by line but are created out of two *interlaced* pictures (fields). One field consists of odd lines numbered 1, 3, 5, 7, etc.; the other field consists of even lines numbered 2, 4, 6, 8, etc., as illustrated in Fig. 2.2. Sending only half of the picture at a time provides reasonable temporal resolution for scenes with motion and reasonable vertical resolution for static images, all within a reasonable bandwidth.

For the watchful eye a horizontal border line seems to flicker in the width of a line. This phenomenon is called *interline flicker.* At a normal viewing distance this line flickering is not very disturbing, and the interlaced transmission has reduced the wide-area flicker that would have resulted from sending the whole picture at half the field rate.

Computer monitors are typically used to view stationary images at a short viewing distance, in which case noninterlaced signals are preferred. Computer screens may be refreshed more than 70 times a second to reduce wide-area flicker.

2.1.3 Resolution

To get an idea of the electrical bandwidth necessary for the transmission of a picture, the number of picture elements to be transmitted must be determined. Given the dimensions and the repetition rate, horizontal, vertical, and temporal resolution can be determined.

There are two temporal resolutions in common TV broadcast use—30-Hz frame or picture rate and 25-Hz frame or picture rate, which correspond to 60-Hz and 50-Hz field rates, respectively. These rates are marginal with respect to wide-area flicker. Beyond 100 pictures per second there is no noticeable enhancement for human beings. For countries using 60-Hz rates, 525 total lines are generated during one frame; for those using 50-Hz rates, 625 total lines are generated during one frame.

Although each line is actually an analog signal, we can use the fact that the aspect ratio of a standard TV picture is 4:3 to calculate that there are about 700 picture elements (pixels) for 60-Hz systems and

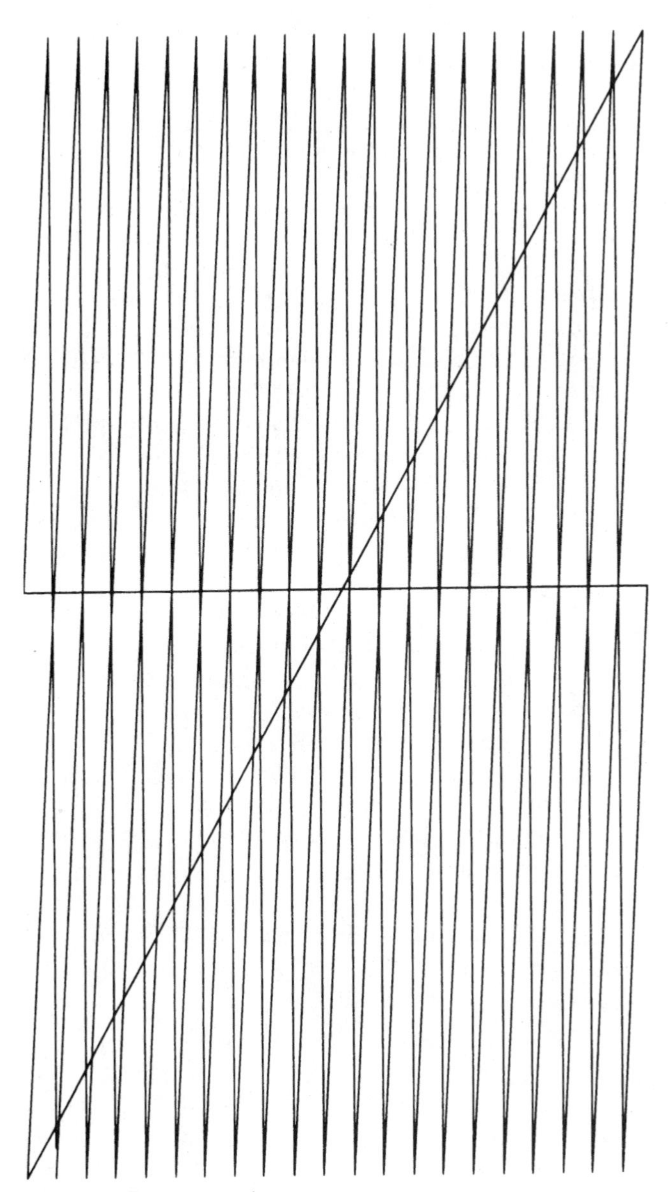

Figure 2.2 Frame scanning pattern.

about 830 pixels for 50-Hz systems. This leads to the time which is available to generate one single pixel:

$$t = \frac{1}{\text{number of frames/s} \times \text{number of lines/frame} \times \text{number of pixels/line}}$$

$= 0.091$ microseconds (μs) for 60-Hz systems or 0.077 μs for 50-Hz systems

The highest video frequency obviously is given when black-and-white pixels are changing rapidly, leading to a period of 0.182 μs for 60-Hz systems and 0.154 μs for 50-Hz systems. Thus, 60-Hz systems need a transmission bandwidth of approximately 5.5 MHz, while 50-Hz systems need about 6.5 MHz. In practice, the bandwidth is reduced to about 4.2 MHz for 60-Hz systems, and to about 5 MHz for 50-Hz systems. Table 2.1 gives an overview of basic system parameters for these two standard black-and-white TV systems.

2.2 Fundamentals of Color

For black-and-white TV it is sufficient to transmit the *brightness.* For a color picture, however, color *hue* and color *intensity* must also be provided. For compatibility reasons, it is necessary that a conventional black-and-white TV set be able to receive a color transmission and reproduce it as a black-and-white picture, and that a color TV set be able to receive and reproduce a conventional black-and-white transmission.

Theories of colors state that nearly every arbitrary color can be composed of three primary colors of appropriate intensity, and the converse is also true, i.e., any color can be split into these three basic colors. In choosing these primary colors it is essential that none of them can be generated from the other two. For color TV, the following three primary colors have been determined:

Color	Wavelength, 10^{-9} m
Red	615
Green	532
Blue	470

TABLE 2.1 Timing for NTSC, PAL, SECAM Television Systems

Field rate, Hz	Lines per frame	Interlace	Line frequency, Hz	Line time, μs	Active video, μs	Bandwidth, MHz
60	525	2:1	15734.266	63.555	52.655	4.2
50	625	2:1	15625.000	64.000	52.000	5.0

Thus, the picture to be transmitted is scanned with three optical devices, each with a different color filter in front of it. The three channels (*R, G,* and *B*) are adjusted such that if a plain white area is scanned, all three outputs have equivalent voltages.

The brightness part of the black-and-white signal is generated by adding the voltages of these primary colors. Because the human eye exhibits different sensitivity for colors of the same intensity, it must be a *weighted* addition. The brightness—or *luminance* (*Y*)—signal is given by

$$Y = 0.299\mathrm{R} + 0.587\mathrm{G} + 0.114\mathrm{B} \tag{2.1}$$

Information about the color *saturation* is already given with the luminance signal; in order to avoid a multiple transmission of saturation, it is subtracted from the color components. These *color difference components* need a lower transmission bandwidth than the luminance information, because the human eye cannot resolve as many color details as brightness details.

Because a specific color hue is determined if two of the three primary colors are known, it is sufficient to transmit information of two color difference signals. The third one can be recalculated at the receiver side.

$$\mathrm{R}-\mathrm{Y} = \mathrm{R}-(0.299\mathrm{R} + 0.587\mathrm{G} + 0.114\mathrm{B})= 0.701\mathrm{R}-0.587\mathrm{G}-0.114\mathrm{B} \tag{2.2}$$

$$\mathrm{B}-\mathrm{Y} = \mathrm{B}-(0.299\mathrm{R} + 0.587\mathrm{G} + 0.114\mathrm{B}) = -0.299\mathrm{R}-0.587\mathrm{G} + 0.886\mathrm{B} \tag{2.3}$$

2.2.1 Compatibility

The luminance signal according to Eq. (2.1) is transmitted just as the black-and-white signal shown previously. The task now is to include additional color information to this signal without losing compatibility.

Fortunately, a black-and-white video picture shows a rather regular spectral density distribution as a result of its scanning nature. Figure 2.3 shows how frequencies within a video band are normally used by the luminance signal. There is concentrated energy around multiples of the line-scanning frequency, and a closer look at a line-frequency bundle shows spectral lines at the distance of multiples of the frame rate. That means there is much spectral space unused by the luminance signal, which can be occupied by the color difference information.

The best choice obviously is to put color information just in the middle of two line-frequency or horizontal bundles of luminance; consid-

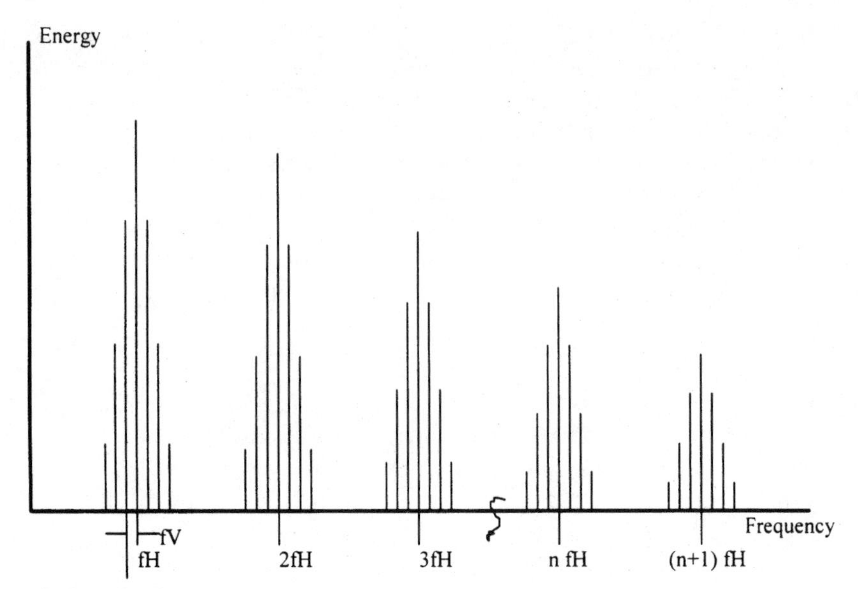

Figure 2.3 Spectral distribution of the luminance signal.

ering also that luminance spectral energy becomes less at higher frequencies (most natural pictures show rather big areas of constant luminance), color difference information should be put into the upper end of the video band.

This can be arranged by modulating the color difference information on an auxiliary subcarrier, the basic principle used by all conventional color-coding standards. The technique of *frequency interleaving* thus needs a subcarrier of an odd multiple of half the line frequency, which is the reason for speaking of a *half-line offset.*

2.3 Color Transmission Standards

The first color transmission standard utilizing the considerations detailed in the previous section was introduced in 1954 in the United States by the National Television System Committee (NTSC). The color subcarrier frequency was chosen as $227.5 \times f_H$ where f_H is the horizontal scanning frequency.

Because *two* kinds of color information, saturation and hue, have to be put on *one* subcarrier, quadrature modulation is applied. Considering the two color difference signals, $(R-Y)$ and $(B-Y)$, as orthogonal vectors spanning a *color plane,* then the length of a composite vector, measured from the origin, is equivalent to color saturation S, while the angle between a composite vector and the x axis [$(B-Y)$ was chosen as the x axis] is equivalent to color hue H.

This can be easily translated into an electrical representation, with the amplitude of the subcarrier corresponding to saturation and the angle of the subcarrier against a certain reference corresponding to hue. So the subcarrier consists of one portion modulated by the $(B-Y)$ component and of a 90° phase-shifted portion modulated by the $(R-Y)$ component. (A color-specific reference signal must also be transmitted, and this signal, called a *color reference burst,* will be discussed later in the chapter.)

Given in mathematical equations, each color (as baseband color difference signals or on a subcarrier) is represented by

$$S = \sqrt{(\mathrm{R}-\mathrm{Y})^2 + (\mathrm{B}-\mathrm{Y})^2\, S} \tag{2.4}$$

$$H = \arctan \frac{\mathrm{R}-\mathrm{Y}}{\mathrm{B}-\mathrm{Y}} \tag{2.5}$$

If the modulated subcarrier is added to the luminance signal, the peak amplitudes of the resulting signals suffer from a ± 80 percent overdrive, compared to the pure black-and-white signal, and this overdrive results in a highly disturbing overmodulation of the transmitter.

In order to avoid this side effect, the color difference signals are reduced by an amount that ensures that the maximum overdrive is limited to ± 33 percent, related to 100 percent luminance (100 percent white) and 0 percent luminance (0 percent black), a number that has proven to be a good compromise.

The reduced $(B-Y)$ signal is called U, while the reduced $(R-Y)$ is called V. Figure 2.4 shows how a color video, blanking, and synchronization (CVBS) signal looks both with and without reduction:

$$U = 0.493\,(\mathrm{B}-\mathrm{Y}) \tag{2.6}$$

$$V = 0.887\,(\mathrm{R}-\mathrm{Y}) \tag{2.7}$$

2.3.1 NTSC encoder

Summing up information given in the preceding paragraphs, a simplified block diagram for an NTSC encoder can be drawn. Figure 2.5 shows a matrix block that translates *RGB* input signals into the *YUV* color space, which obviously is more appropriate for transmission. Both the luminance signal and the color difference signal are bandlimited in accordance to broadcast specifications.

An NTSC-specific item is the I and Q representation of the U and V signals. Many examinations of human visual perception have shown that the eye not only is more sensitive for brightness details but also

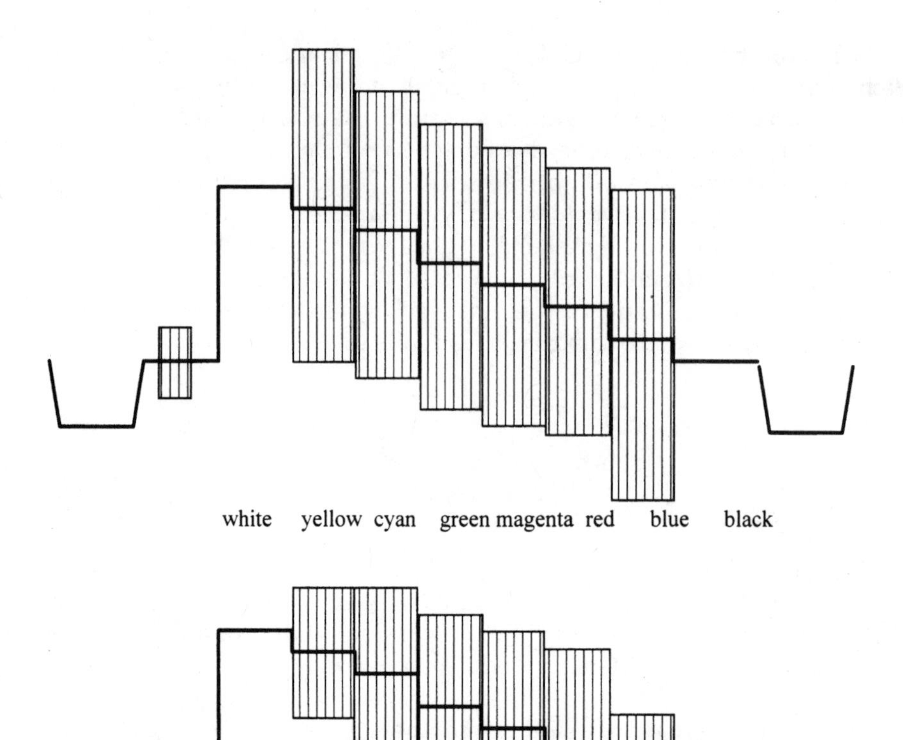

Figure 2.4 Scan line: (top) without reduction, (bottom) with reduction.

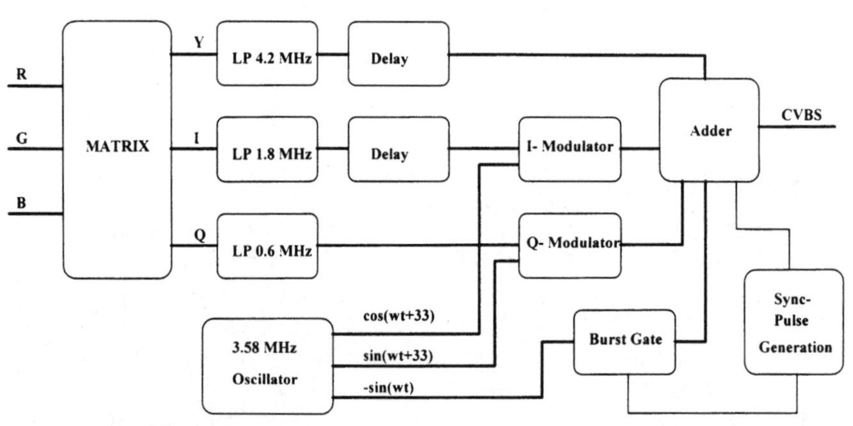

Figure 2.5 An NTSC color encoder.

that there are differences with respect to color hue sensitivity. Color steps going from blue to yellowish green are perceived less clearly than color steps going from orange to cyan. Thus, less bandwidth is necessary for color difference signals in the blue-green domain than in the orange-cyan domain. The NTSC encoder makes use of this effect, spending about 0.6 MHz and 1.8 MHz for the Q and I signals, respectively.

Because the I and Q axes are not identical to the V and U axes, respectively, the U and V signals have to be translated into the $I-Q$ color space (the angle between the U and Q axes equals 33°):

$$I = V\cos 33° - U\sin 33° \tag{2.8}$$

$$Q = V\sin 33° + U\cos 33° \tag{2.9}$$

An additional delay line is needed in the luminance path because the more band-limited color difference signals introduce more propagation delay. In an NTSC $I-Q$ system, even the I portion needs an extra delay compensation. The reference signal necessary for the color decoding process, the burst, consists of about nine periods of subcarrier. It follows the horizontal synchronization pulse, controlled by a gating pulse from the timing master. Its phase relationship within the color vector plane is always the $-(B-Y)$ axis.

2.3.2 NTSC decoder

The main task of the decoder is to regenerate the saturation and hue of the encoded color difference signals and to separate them from the luminance, which is embedded in the same transmission channel.

Because hue information is coded into the phase of the modulated subcarrier relative to the burst phase, it is essential to reconstruct the phase of the demodulating oscillator out of the burst. The spectral energy of the burst, however, is spread very widely, because its pulse duty cycle is only about 2.5:64. A *phase-locked loop* (PLL) of rather small bandwidth is necessary for reliable demodulation with a quadrature demodulator corresponding to the quadrature modulator on the transmitter side. The locking range of the PLL must not exceed $\pm f_H/2$, because otherwise it might lock to a side band of the burst rather than to the main spectral component.

Saturation is coded into the amplitude of the modulated subcarrier relative to the burst amplitude. Thus, burst amplitude can also be used for an automatic gain control of color, because different amounts of attenuation of luminance (which is calibrated to the amplitude of synchronization pulses) and chrominance can occur in transmission.

2.3.3 PAL

The phase alternating line (PAL) was developed as a consequence of NTSC, avoiding the disadvantage of wrong color hue due to phase errors of the quadrature-modulated color difference signals during transmission or during coding or decoding. Because the color synchronization signal, the burst, is always built up on a constant voltage level and since the color difference is on a continuously moving luminance level, the phase relation of color difference to burst can vary.

As long as such phase errors are constant over a long period, they are canceled out by the feedback loop of the demodulating oscillator in the receiver. However, short time errors caused by high-frequency luminance lead to differential phase errors, resulting in visible hue distortions. The PAL compensates for such errors, if

- The phase error of two consecutive lines is nearly the same
- The color saturation and hue of two consecutive lines are nearly the same

Figure 2.6 illustrates the principle. The phase error of a color vector from its nominal position is compensated by inverting the $(R-Y)$ component in every second line. At the receiver, this inversion is removed again after averaging two lines.

Additionally, in order to identify in which line the $(R-Y)$ inversion was done in the PAL encoder, the color burst is switched line by line with respect to its $(R-Y)$ portion. Compared to the NTSC encoder, the nominal phase angle of the burst is 135° in the noninverted line and 225° in the inverted line.

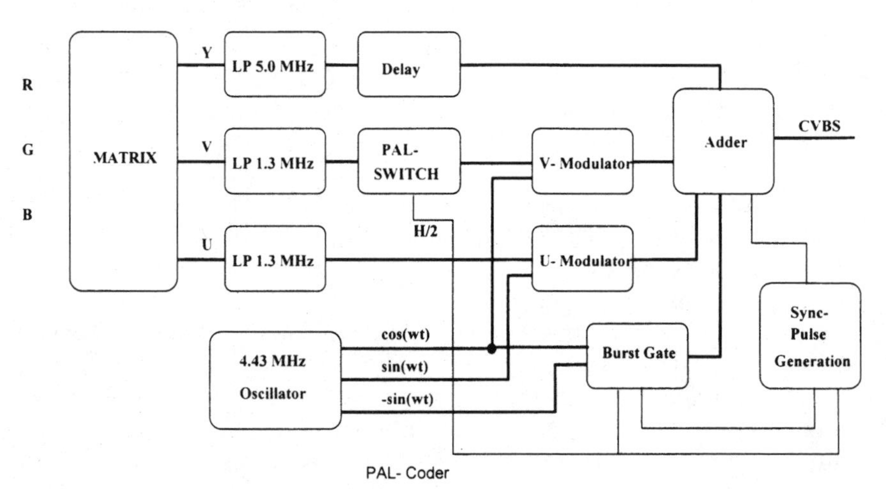

Figure 2.6 A PAL color encoder.

2.3.4 PAL encoder

An examination of the PAL encoder shown in Fig. 2.6 reveals blocks similar to those of an NTSC encoder. Note, however, the polarity switch for the $(R-Y)$ signal and burst according to the previous considerations. The U and V components of equal bandwidth are modulated directly on the subcarrier (without conversion into $I-Q$ color space). Line-by-line inverting of the $(R-Y)$ component introduces an additional periodicity of half the line frequency, thus modifying the spectrum of the modulated color, as shown in Fig. 2.3.

If a half-line offset of the subcarrier was chosen, as with the NTSC encoder, then portions of the luminance spectrum would fall on modulated $(R-Y)$ portions, and correct demodulation could not be done. To suppress moving interference patterns, the unmodulated subcarrier frequency is additionally shifted by an amount corresponding to the frame rate. For PAL systems, a quarter-line offset is used; thus for the conventional PAL encoder the subcarrier frequency equals $283.7516 \times f_H = 4.43361875$ MHz.

2.3.5 PAL decoder

To illustrate the mechanism of eliminating differential phase errors with the PAL, an example of a disturbed and recovered vector is given in Fig. 2.7. The initial angle at the coder may be φ_1. Because of a differential phase distortion, a resulting phase of

$$\varphi_{\text{res1}} = \varphi_1 + \varphi_d$$

is detected at the decoder. In the following line, $(R-Y)$ is inverted; thus

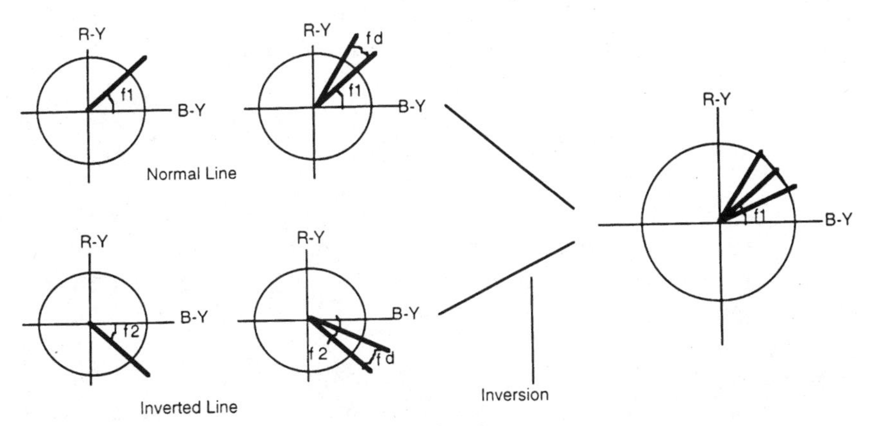

Figure 2.7 Vector representations of a PAL decoder.

$$\varphi_2 = 360° - \varphi_1$$

is generated, and

$$\varphi_{res2} = \varphi_2 - \varphi_d$$

is detected. In the receiver, the $(R-Y)$ component of the second line is reinverted, resulting in

$$\varphi_{res2}{}^* = 360° - (360° - \varphi_1 + \varphi_d) = \varphi_1 - \varphi_d$$

When both vectors of the two consecutive lines are added geometrically, the phase error φ_d disappears, and the resulting angle is φ_1.

For averaging two lines, a line memory is necessary, which is a characterizing part of a conventional PAL decoder. The subcarrier PLL needed to demodulate the color can be considered to be similar to the needs for the NTSC decoder, but it can be seen that because of the tighter spectrum, side lock is even more dangerous.

2.3.6 SECAM

Sequentiel Couleur Avec Memoire (SECAM), developed in France, addresses the problem of controlling the vertical resolution of color. Because color information of two consecutive lines usually does not differ very much, $(B-Y)$ and $(R-Y)$ can be transmitted sequentially, so only one single color difference signal is apparent at a time. At the receiver, again as with the PAL, a line memory is needed in order to store one of the two color difference signals such that both signals are available simultaneously.

2.3.7 SECAM encoder

A simplified block diagram of a SECAM coder is shown in Fig. 2.8. Color difference signals D_b and D_r are normalized according to

$$D_b = 1.505\,(B-Y) \tag{2.10}$$

$$D_r = -1.902\,(R-Y) \tag{2.11}$$

The difference factors in Eqs. (2.10) and (2.11) are the result of unequal amplitude distributions of the color difference signals. This is important, because modulation in SECAM is *frequency* modulation (FM) instead of *amplitude* modulation (AM). Therefore a preemphasis, in combination with a clipping circuit to protect against overmodulation, is used before the signals are frequency-modulated line by line on their respective carriers. The transfer characteristic of the

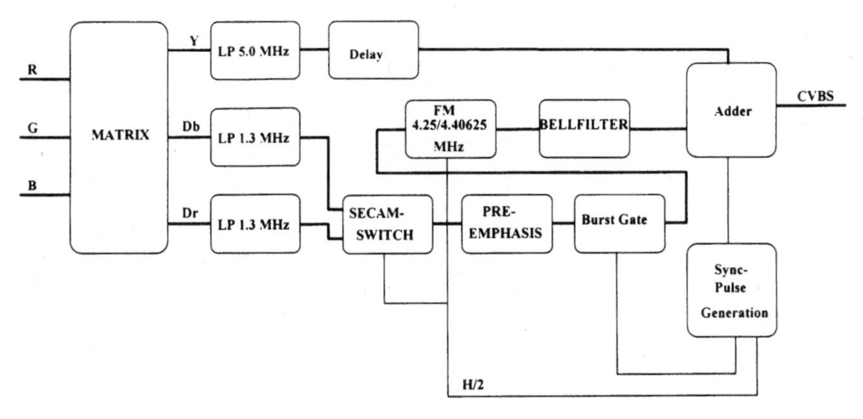

Figure 2.8 A SECAM color encoder.

preemphasis $H(f)$ is expressed by

$$H(f) = \frac{1 + j(f/f_1)}{1 + j(f/3f_1)} \tag{2.12}$$

where f is the signal in kilohertz (kHz) and $f_1 = 85$ kHz.

Carrier frequencies are chosen as

$$F_{0b} = 272\, f_H = 4.25000 \text{ MHz}$$

$$F_{0r} = 282\, f_H = 4.40625 \text{ MHz}$$

The deviation corresponding to the color saturation after preemphasis must be asymmetrical, since otherwise too much bandwidth would be occupied. (See Table 2.2.)

With frequency modulation, the subcarrier is continuously apparent with low-saturated colors, where it is most visible. So an additional amplitude modulation is applied, such that at low saturation (corresponding to low deviations), the carrier amplitude is reduced relative to high saturation.

Amplitude modulation is done by feeding the FM signal into the so-called *bell filter*, that characteristic of which is described by

TABLE 2.2 Deviation as a Function of Signal Type

Signal	Deviation static, kHz	Deviation dynamic, kHz
D_b	± 230	+ 506/−350
D_r	± 280	+ 350/−506

$$H(f) = G\frac{1 + j16F}{1 + j1.26\,F} \tag{2.13}$$

with $F = (f/f_0 - f_0/f)$
$f_0 = 4.286$ MHz
$G =$ constant

As in PAL transmissions, an identification signal is provided with a SECAM burst signal that contains the unmodulated F_{0b} or F_{0r} subcarrier corresponding to the actually transmitted color difference signal.

2.3.8 SECAM decoder

All SECAM-specific blocks of the coder have to be incorporated with their inverse function in a SECAM decoder, i.e., after an *anti–bell filter* frequency demodulation with an appropriate FM demodulator and deemphasis of the baseband signals. A line memory is necessary to store color information of the preceding line for displaying it together with color information of the current line.

2.3.9 Pros and cons of the three main standards

Comparing NTSC, PAL, and SECAM standards, we can find inherent advantages and disadvantages in each. The NTSC standard theoretically delivers the best color reproduction, as it shows no loss of vertical color resolution, and, if $I-Q$ modulation is applied, also gives the best horizontal color resolution. On the other hand, it is the most sensitive to phase errors, resulting in inaccurate color tints at the displaying TV set. The frequency of the color subcarrier is rather low; i.e., these frequencies are close to perceivable frequencies of luminance, which often leads to cross distortions. Such distortions may appear as dots in the luminance during sharp color transitions, and also as highly saturated color flickering in sharp-edged luminance components.

The PAL standard is rather insensitive to phase errors but suffers from a certain loss of vertical resolution because of the line-averaging technique. Cross distortions are not as critical as with NTSC because PAL employs a higher subcarrier frequency, but some cross distortions are still apparent.

Finally, the SECAM standard is least sensitive to hue distortions, because the color information is transmitted line by line with frequency modulation. This leads directly to its primary disadvantage, which is an annoying line flickering at 12.5 Hz on sharp vertical color transitions. With poor input signals, a SECAM decoder shows significant color stripes resulting from low subcarrier amplitudes which the demodulator misinterprets as zero crossings. These singular demodulating errors are stretched to stripes (called *fish*) by the deemphasis. Clipping of the coder cannot be redone in the decoder, so sharp color transients cannot be reproduced as well as with NTSC or PAL.

2.4 Special Considerations in Digital Processing of Video Signals

In processing digital video, the effect of jitter of the sampling clock plays an important role with respect to picture quality. *Jitter* is uncertainty of the exact instant when the sample will be taken. This can degrade, for example, the quality of a color demodulation process or the stability of a displayed picture element on a cathode-ray tube. Jitter is particularly critical during analog-to-digital conversion, because unstable sampling points on analog signals give a distorted digitized representation, regardless of the linearity performance of the ADC itself.

The effect of this jitter (also called *aperture uncertainty*) can be treated as additional noise that is superimposed onto the ideal result of conversion (Fig. 2.9.) With an aperture uncertainty of $\Delta t(t)$ we can formulate

$$F[t + \Delta t(t)] \approx f(t) + f'(t)\,\Delta t(t)$$

with $f'(t)\,\Delta t(t)$ describing the error due to jitter.

The mean value of this noise is given by

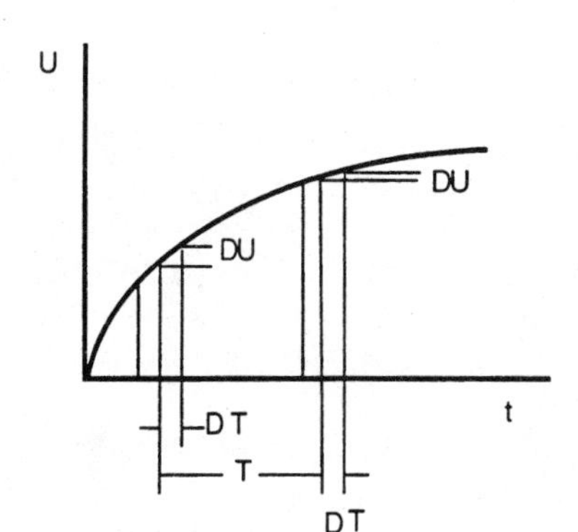

Figure 2.9 Sampling jitter in an ADC.

$$\overline{[f(t) \times \Delta t(t)]^2} = \overline{[f(t)]^2} \times \overline{[\Delta t(t)]^2}$$

The root mean square (rms) value of this noise is the square root. Assuming a sinusoidal voltage of frequency f and amplitude A is applied to the input of the ADC, the effective jitter voltage is given by

$$V_{\text{jitter (rms)}} = \sqrt{\overline{\left[A\,2\pi f \cos(2\pi ft)\right]^2}\left(\overline{t_{\text{rms}}}\right)^2} = 2\pi\, Af\Delta t_{\text{rms}}$$

It is clear that the peak error of jitter increases in direct proportion to the first derivative of the sampled signal. So, in the case of a sine wave, the zero crossings are most sensitive to jitter of the sampling clock.

2.4.1 Noise

It is significant that for a digitized signal all samples are discrete numbers within a limited range. With n-bit resolution, there are $2n$ quantizing intervals. An important consequence is that the code of a digitized signal (a signal that was converted from analog to digital) cannot change until the analog input signal changes by the amount of $\pm$ 0.5 of the least significant bit (LSB), measured from the middle of the quantizing interval. The largest quantizing errors occur as these 0.5 decibel (dB) limits are approached. Therefore, it is not possible to reconstruct out of a digital code the exact original analog value by means of an ideal DAC.

The effect of quantizing errors is also described as *quantizing noise,* the level of which is directly dependent on the actual digital resolution. This quantizing error can be treated in general as a random value, and its distribution can be described by a probability density function, as long as the signals to be quantized are distributed statistically. Then the probability for a quantizing error is equally distributed between $+q/2$ and $-q/2$ (where q = 1 LSB).

In order to determine the quantization noise at a given number of bits, an ideal ADC may generate a digitized ramp signal, which is converted into an analog ramp again by a subsequent DAC. By comparing the output with the input, the quantizing error can be measured, which appears as a sawtooth voltage between $+0.5q$ and $-0.5q$ (Fig. 2.10). The root mean square of this sawtooth signal is given by

$$\text{err}_{\text{rms}} = \frac{1}{T}\int_{-T/2}^{+T/2}\left(\frac{q}{T}\times t\right)^2 dt = \frac{q}{\sqrt{12}}$$

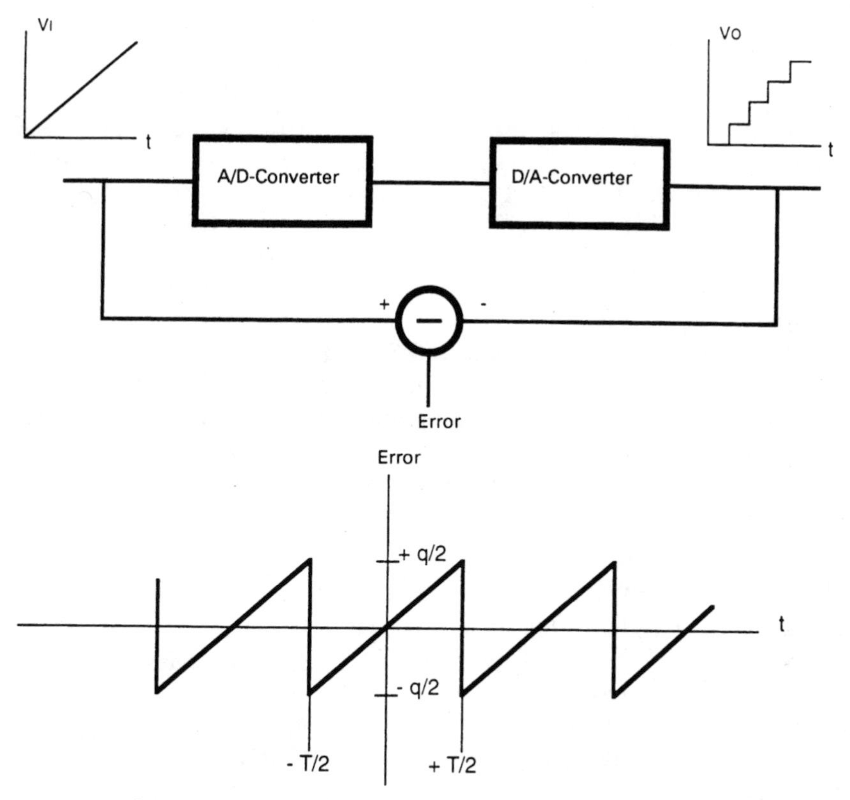

Figure 2.10 Quantization noise characterization using a sawtooth waveform.

To convert a sine wave of type $U \sin x$ using n bits, $2^n - 1$ steps can be used, resulting in a root mean square of

$$U_{\text{rms}} = 2^n - 1 \times \frac{q}{\sqrt{2}}$$

Given the signal-to-noise ratio (SNR) with

$$\frac{S}{N} = \frac{U_{\text{rms}}}{\text{err}_{\text{rms}}}$$

it can be calculated to be

$$2^{n-1}\sqrt{\frac{12}{2}} = \sqrt{1.5} \times 2^n$$

or in logarithmic representation as

$$S\backslash N\ dB = 6.02n + 1.76$$

The SNR thus increases by 6 dB with each additional bit of resolution.

For video signals, it is common to define the SNR by comparing peak-to-peak video to rms noise, resulting in about + 9 dB for a given resolution. For a video SNR of 60 dB, at least 8 bits are required, assuming that a full-scale video signal occupies the full digital word range.

Because quantization noise in general is equally distributed throughout the *Nyquist band* (the frequency range between zero and half of the sampling clock), the initial SNR after analog-to-digital conversion can be enhanced by oversampling and successive filtering. As each filter stage reduces the signal bandwidth by a factor of 2, it also reduces noise energy by a factor of 2, thus by gaining better resolution. It should be noted that not all types of filters produce such bit savings, such as simple averaging filters.

2.4.2 Special filter techniques

Whenever the bandwidth of video signals must be adapted for particular purposes, filters—such as low-pass filters, bandpass filters, or notches—become necessary. For example, in a common color decoder, a notch filter is used to suppress the color subcarrier in the luminance channel at approximately 4.43 or 3.58 MHz. A low-pass filter applied after the demodulation process in the chrominance channel is also necessary in order to suppress the mirror signal of two times the subcarrier frequency. When digital signal processing is used, such filters can be treated as *moving averagers* in the horizontal picture axis, because they combine pixels which are immediate neighbors.

For example, a steep black-to-white transition can be converted to a smoothed black-to-white transition with a 50 percent gray value by using a circuit combining two neighboring pixels to an average value, thus reducing the bandwidth of the input signal. This approach is illustrated in Fig. 2.11.

$$\text{Output}(n) = \frac{\text{input}(n) + \text{input}(n - 1)}{2}$$

Obviously, in this simple example pixel storage is needed that delays a pixel by the duration of one sample period.

For delaying a sample for the duration of one picture line (as needed in a common PAL decoder), a memory capable of storing a complete video line is required. When long delays are needed for complex filtering, digital signal processing offers a significant advantage over analog processing.

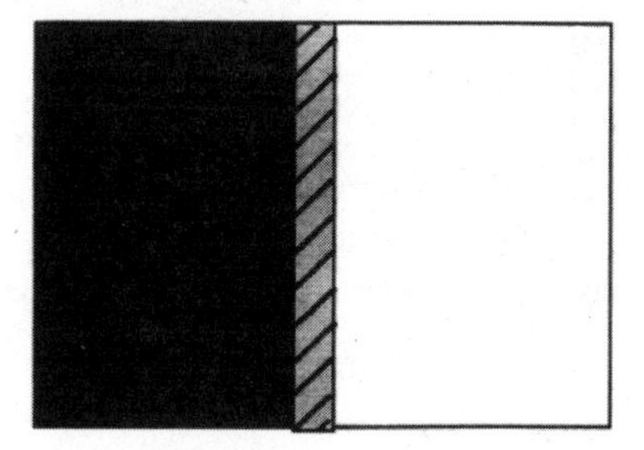

Figure 2.11 Low-pass filtering of a transition from black to white.

2.4.3 Spatial filtering

Averaging in the vertical direction with delay lines is also required when the size of video images is changed. For example, reducing the size of a picture means less pixels in the horizontal direction and less picture lines in the vertical direction, thus dropping a certain amount of pixels and complete lines. Because the original resolution of the picture has been reduced, the bandwidth of the signal must be reduced in order to avoid aliasing due to subsampling. In this case, the averager would not only produce smooth transitions in the x axis, but also in the y axis, before pixels and lines are dropped for scaling (Fig. 2.12).

2.4.4 Temporal filtering

The last step in multidimensional filtering is related to the time axis. As a first example, field-rate conversion of a video picture is considered. Field-rate conversion, for example from 50 to 100 Hz, eliminates disturbing wide-area flicker by doubling the field rate. This is achieved by storing a complete field in a digital memory and successively reading each stored field twice at double the original speed.

A second large area of temporal filtering, combined with spatial filtering, is associated with noise reduction. Averaging consecutive pic-

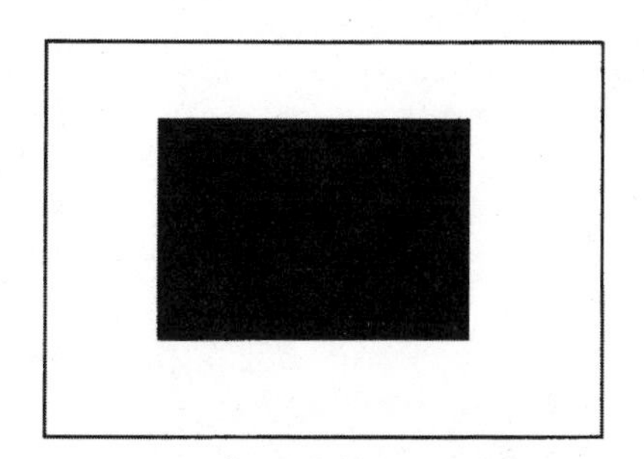
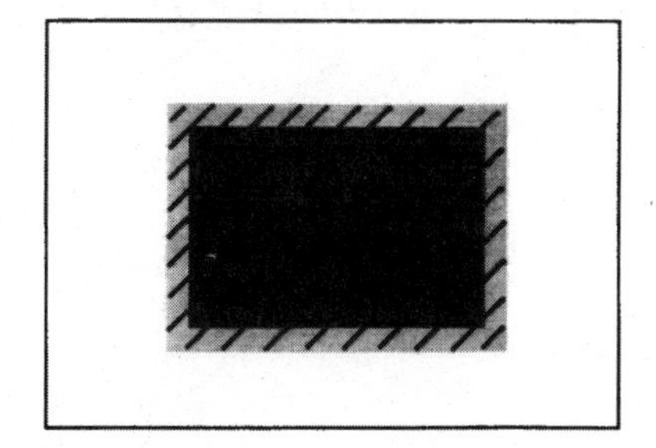

Figure 2.12 Spatial filtering of a black-to-white transition.

tures also averages noise, which, because of its random character, is canceled out. Since the spatial resolution of the picture is only slightly reduced, the result is a noiseless but sharp picture.

Finally, in the case of video compression, some systems such as MPEG need temporal interpolation filters to generate intermediate pictures, leading to a temporal resolution corresponding to the uncompressed original data stream.

Chapter

3

Video Compression Techniques: Coding

3.1 The Picture as a Message

The discussion of the human visual system in Chap. 1 emphasized that the transmission of a picture sequence does not have to be done without any errors because many errors will not be perceived by the viewer. Thus, it may be possible to reduce bit count without perceivable problems; this particular procedure is called *reduction of irrelevance.* The picture content itself offers another starting point for data compression. Pictures normally contain redundant information, and this chapter deals with the picture itself as the source of the redundant information.

The term *codec* will be used to describe a system including a coder and a decoder. This chapter provides an approximate reference model for a codec and an explanation of its basic components. It also will discuss simple models for the picture signal and examine how such models influence coder design. It was the very useful idea of C. E. Shannon[1] to decouple the problem of information transfer from the information to be transferred, that is, the picture content. In this way it is possible to mathematically model the information transfer and even to express a message in such a thing as bits. Applied to picture communication, this means that the transmitted bits do not necessarily need to have anything to do with the picture content. The receiver of the transmitted picture—usually a frame, if it is one picture from a sequence of pictures—will choose from a stock of all possible pictures.

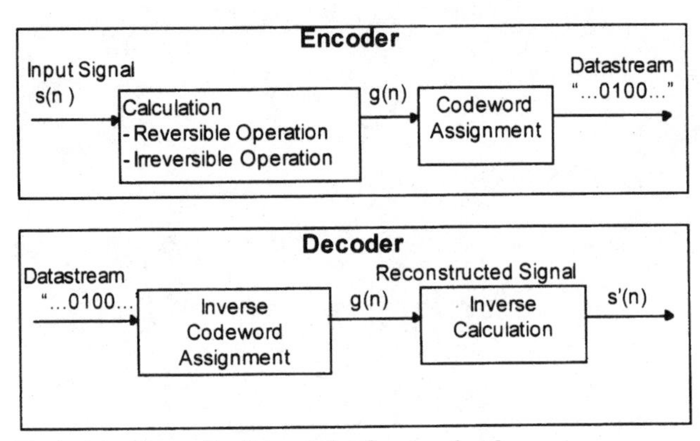

Figure 3.1 Generalized example of a signal codec.

3.2 Codec Reference Model

Figure 3.1 shows a rough reference model of a codec in the signal stream. The input signal is a succession of picture elements $s(1)$, $s(2)$, $s(3)$,..., $s(n)$. The picture signal $s(n)$ is processed by several reversible and irreversible operations carried out by a calculation unit. A unique code word must be attached to the various amplitudes of the signal $g(n)$ so that it can be clearly reproduced at the decoder. This code-word designation can, for example, be realized as a code-word table in which the code words can have differing lengths. The incoming code word will be decoded at the decoder with the same code-word table, and the resulting signal value $g(n)$ will be supplied to the inverse calculation unit. The output signal of this expression should, from a subjective point of view, represent the input signal of the coder. The exact matching of the input and output signal is, of course, only possible when the specific calculation unit of the coder does not contain any irreversible operations. If we want to analyze or design a codec, we need to look again at the primary aim of picture coding: the representation of the picture signal with as few bits as possible. These bits are nothing other than the codes read out of the code-word table. With the systematic design of the calculation unit and code-word assignment, we are basically faced with two questions:

1. What characteristics does the input signal $g(n)$ of the code-word assignment need to have so that a code-word table can be found which, on average, produces the smallest number of bits? The following factors can play a role here:

   - The number of the different values that $g(n)$ can obtain

- The frequency or the corresponding probability of appearance with which the values appear
- The sample rate of $g(n)$

2. How must the calculation unit be designed relative to the characteristics of the input signal $s(n)$ so that the input signal of the code-word assignment $g(n)$ has the characteristics required in question 1?

Some examples of calculation units and code-word assignments have already been touched on. For example, performing a transformation and sending the difference of successive (in time) frames are reversible operations. The representation with fewer gray levels, i.e., the mapping of several gray levels to a corresponding average gray level, is surely an irreversible operation, as is the suppression of signal frequencies using low-pass or high-pass filtering algorithms. An example of a simple though not necessarily suitable code-word assignment is the coding of a gray level with the binary representation of its amplitude (for example, the number 5 is represented by 0000 0101 in 8-bit straight binary code). Figure 3.2 gives a simplified assignment table using four gray levels.

3.2.1 First picture model and its coding

Let us assume that the encoder is comprised only of a code-word assignment block (as in Fig. 3.2). How then can a suitable coding table for the gray levels of the picture be designed? How can we judge whether a given code is optimal? Information theory offers few useful tools for answering these questions. In order to be able to apply these tools to the picture material, the picture material itself must be modeled in terms of information theory. We therefore make the following models (Fig. 3.3):

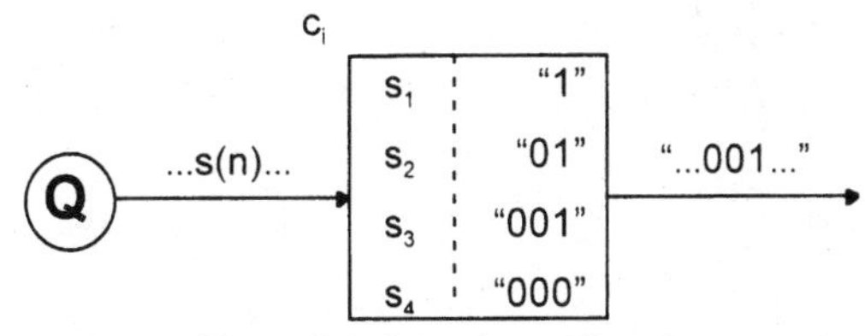

Figure 3.2 Example of a code-word assignment for four gray levels.

Reality: Brightness values scanned horizontally across a line

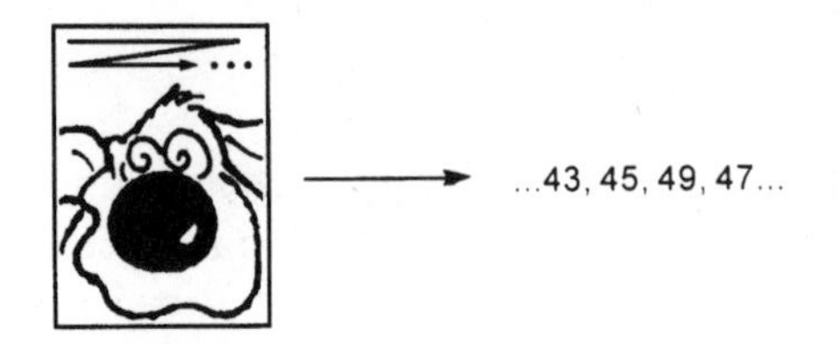

Assumed Model:

1. Fixed source alphabet

Q → ...s(n), s(n+1), s(n+2)...

$s(n) \in \{s_1, s_2 ... s_{256}\} = \{\text{"0" "1"..."255"}\}$

2. Fixed probability of events

s_1 - $P(s_1)$, s_2 - $P(s_2)$, s_N - $P(s_N)$

Figure 3.3 Modeling of a picture signal.

1. The picture material constitutes an information source which produces, successively, source events $s(n-1)$, $s(n)$, $s(n+1)$,...out of a fixed source alphabet A, where $s(n)$ is an element of A and $A = \{s_1, s_2, \ldots, s_N\}$. The source events $s(n)$ can, for example, be the gray levels that are read out of the digitized frame, pixel for pixel. With 256 gray levels, the source alphabet $A = \{0, 1, —, 255\}$ is composed of elements s_i and has an index $i(s) = 1, 2, —, N$.

2. All source events have a fixed probability of appearance $P(s_i)$. If, for example, the gray level $s(n) = 130$ occurs in 2 percent of all cases, the probability $P(130) = 0.02$.

In order to introduce the code-word assignment into the model, the terms *code-word length* and *average code-word length* are defined:

- The code-word length l_i describes the number of bits of code word C_i, with which the source event $s(n) = s_i$ is encoded. If we encode, for example, s_3 with $c_3 = 001$, then $l_3 = 3$ (as in Fig. 3.2).
- The average code-word length L describes the number of bits that are transmitted per symbol and is calculated as a result of a long-term observation

$$L = \sum_{i=1}^{N} P(s_i) l_i$$

As an experiment, if all l_i's are equal, then $l_i = l$, gives

$$L = \sum_{i=1}^{N} P(s_i) = l \sum_{i=1}^{N} P(s_i) = 1$$

An interesting relationship exists for the average code-word length L, which, if the model is appropriate, also determines the average bit rate. Namely, we can prove:

$$L \geq H \qquad \text{with } H = -\sum_{i=1}^{N} P(s_i) \log_2 P(s_i)$$

As a reminder for those of you who left high school a long time ago,

$$\log_2 2^K = K$$

There is also a lower limit H for the bit rate, which is only dependent on the probability of appearance $P(s_i)$ of the source. The value H is called *entropy* or *information content* of the source. It has a central significance in the information theory which is: There is no code for the information source which makes the average code-word length smaller than the entropy H of the source. Without having devised a code, we can estimate, through simple calculation of the entropy, the average code-word length per source event (or average data rate) that will occur. The further a code approaches entropy H with its average code-word length, the greater is its so-called *coding efficiency,* and the smaller is the *code redundancy.*

The entropy has some notable characteristics, which are shown in Fig. 3.4. For an information source with N different events, the

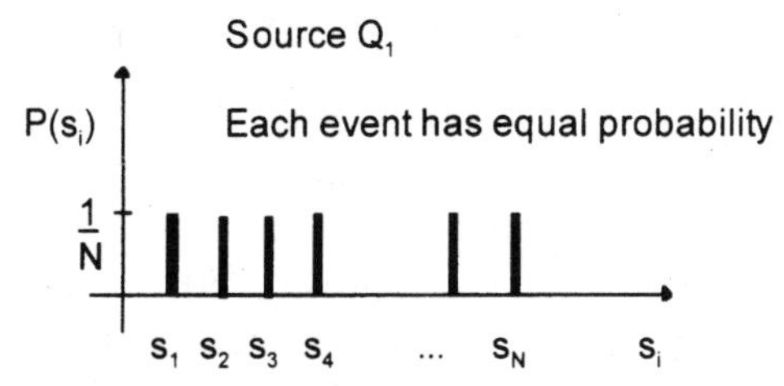

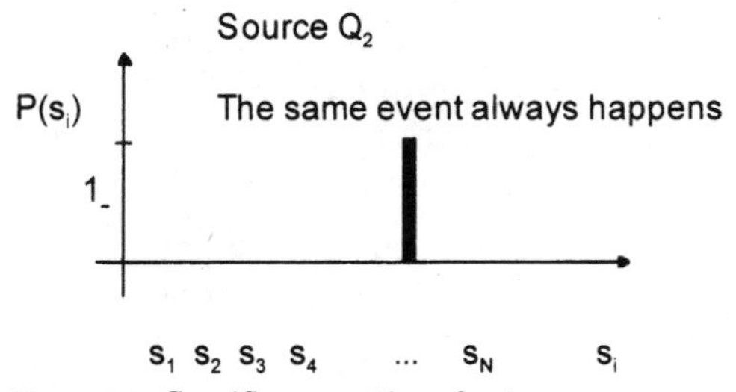

Figure 3.4 Specific properties of entropy.

entropy will be greatest when $H = \log_2 N$, when on average, all events occur with the same regularity. The more irregularly the probabilities are distributed, the smaller will be the value of the entropy. If in the most extreme case, only one event from N possible events occurs, the entropy H becomes $H = 0$. If we consider a picture digitized with 256 gray levels, which on average all occur with the same regularity, we use at least $H = \log_2 256$ bits $= 8$ bits to code a pixel. If, in the other extreme, only one gray level occurs, the entropy H will be zero. In fact no data need to be transmitted in this case because the signal of this source, namely, the single gray level, is known already. Some typical entropy values for three different sources are as follows:

- The difference picture of a sequence of frames has the smallest entropy of $H = 3$ bits because of its prominent distribution density; only a few small amplitudes occur, but these occur frequently.
- The corresponding typical picture has an entropy of about $H = 7$ bits if the preceding and successive pictures or frames are not taken into account.
- A smooth ramp has the largest entropy, with a linear increase in brightness from left to right. All gray levels occur there with the same regularity. Nevertheless, it is hardly credible that a "boring" ramp contains more information than a natural picture. Here we have the first limitation of the picture model.

3.2.2 Code-word assignment for a particular source

An average code-word length which is at the theoretical lower limit may turn out to be useful, but the problem of finding a suitable code-word allocation is, unfortunately, not yet solved. There are various algorithms for designing encoding tables. The most famous—Huffman's—was introduced in 1952 and still proves to generate the best code.[2,3] The recursive algorithm for generating the code is shown in Fig. 3.5. It has two steps.

- *Step 1: Summarize and sort the probabilities of appearance.* The two smallest of all the probabilities are summarized from left to right, and the results are always inserted into the remaining probabilities. At the end, only two probabilities remain.
- *Step 2: Designate the bits.* The two remaining probabilities are designated as 0 and 1, respectively. The summarized probabilities are now consecutively split up into their original values, a 0 being designated to the upper branch and a 1 to the lower branch. This

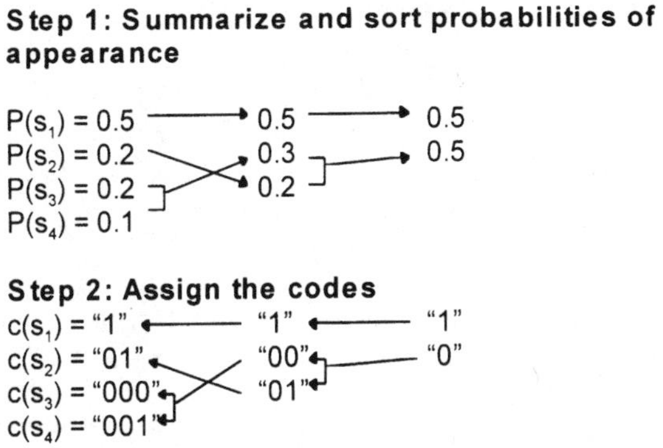

Step 1: Summarize and sort probabilities of appearance

$P(s_1) = 0.5$ → 0.5 → 0.5
$P(s_2) = 0.2$ → 0.3 → 0.5
$P(s_3) = 0.2$ → 0.2
$P(s_4) = 0.1$

Step 2: Assign the codes

$c(s_1)$ = "1" ← "1" ← "1"
$c(s_2)$ = "01" ← "00" ← "0"
$c(s_3)$ = "000" ← "01"
$c(s_4)$ = "001"

Figure 3.5 Design of a Huffman coder for a source with four elements.

algorithm generates a code of which the average word length cannot be larger than H+1; thus:

$$H \leq L \leq H + 1$$

The effectiveness of the Huffman code is only reduced slightly by small deviations in the occurrence distribution.[3] There are nevertheless methods of code-word assignment that even allow for adjustment to the momentary statistics of the signal to be coded but which have other disadvantages.[4–6] It should be noted that within a Huffman code table, no code word can be the beginning of a longer code word. The decoder, therefore, knows with which bit a transmitted code word ends and with which the next code word begins in the event of a transmission error. Following are two experiments with the application of entropy and Huffman coding.

3.2.2.1 Experiment 1: Coding of a black-and-white piece of text. Assume that a piece of text needs to be coded in which only black or white pixels occur (Fig. 3.6). The black pixel appears in 20 percent and the white in 80 percent of all cases. Thus

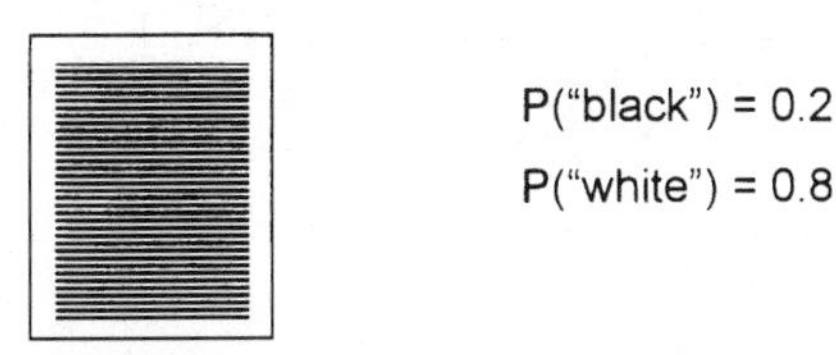

Figure 3.6 Coding of a black-and-white piece of text.

$$A = (\text{black, white})$$

$$P(\text{black}) = 0.2 \qquad P(\text{white}) = 0.8$$

The entropy of this source is therefore

$$H = -(0.2\times\log_2 0.2 + 0.8\times\log_2 0.8)\ \text{bit/pixel} = 0.72\ \text{bit/pixel}$$

We know therefore that, on average, a minimum of 0.72 bits/pixel is required for coding. The generation of a Huffman code is trivial: one of the two events must be coded with a 1 and the other with a 0. The average code length would therefore be

$$L = 1\ \text{bit/pixel}$$

As given by the Huffman code, $L \leq H+1$. This is, however, only a small consolation that the rise in data by 0.28 to 1 bit/pixel is higher than the information H. This is nevertheless 39 percent more data than a transmission with 0.72 bit/pixel.

3.2.2.2 Experiment 2: Extension of the source. The next example shows that, although the entropy only represents an unachievable lower limit, it can in fact be approached fairly closely. If we take the same piece of text as above, two consecutive pixels are coded with a code word (Fig. 3.7), thus giving a new source with four events, namely,

$$A = (w, w; w, b; b, w; b, b)$$

where b = black and w = white.

The probability of occurrence for each of the pixel pairs is derived from the product of the probability of the pixels contained in the pixel pair. We therefore obtain the probability for all four events and can once again devise a Huffman code:

$$P(w, w) = 0.64 \rightarrow c_1 = 1 \qquad I_1 = 1$$

$$P(w, b) = 0.16 \rightarrow c_2 = 01 \qquad I_2 = 2$$

$$P(b, w) = 0.16 \rightarrow c_3 = 001 \qquad I_3 = 3$$

Figure 3.7 Simultaneous coding of two pixels.

$$P(b, b) = 0.04 \rightarrow c_4 \quad = 000 \quad I_4 = 3$$

For entropy H' and average code-word length L' of a pair of pixels, this gives

$$H' = [0.64 \times \log_2 0.64 + 2 \times (0.16 \times \log_2 0.16) + 0.04 \times \log_2 0.04] \; bit/(2\,pixels)$$

$$= 1.44 \text{ bit/(2 pixels)}$$

$$L' = [1 \times 0.64 + 2 \times 0.16 + 3 \times 0.16 + 3 \times 0.04] \text{ bit/(2 pixel)} = 1.56 \text{ bit/(2 pixels)}$$

For entropy H, and the corresponding code-word length L for one pixel, the following applies:

$$H = \frac{H'}{2} = 0.72 \text{ bit/pixel}$$

$$L = \frac{L'}{2} = 0.78 \text{ bit/pixel}$$

We can make two interesting observations here:

- The entropy H' for the pixel pairs is double that of the entropy per pixel in the first example. As was to be expected, the information content per pixel has stayed the same because it was dependent on the source and not on the way in which it was coded.
- As before, for the Huffman code, the average code-word length L' for a pixel pair has an upper limit due to H'+1. Therefore, the average code-word length L for a single pixel is only limited by $(H'+1)/2$ = $H+\frac{1}{2}$. The average code-word length per pixel (L = 0.78 bit/pixel) has, in fact, moved closer to the entropy at 0.72 bit/pixel.

The summary of n events during coding is called *extension of a source by factor* n. The previous example was the extension by factor 2 of the source (black, white). If we develop the consideration further, it becomes clear that extension by factor n limits the average code-word length L to $H<L<H+1/n$. The closeness of L to entropy H can therefore be influenced by the suitable selection of n. Hence, it is possible to code an information source with a data rate that is as close as required to the information content of the source, known as the *first Shannon theorem.*

3.2.3 Second picture model: The source remembers

Unfortunately, during the modeling of the source no consideration was made of the fact that source events can be dependent on each other. If, for example, we take a piece of text, it is evident that there are many closed white and black areas. The probability of being able to find a black pixel next to another black pixel can be higher than that of being able to find a white pixel next to the black pixel, despite white being, in totality, present more often. Naturally, we also observe similarities between neighboring pixels. These considerations lead to a new source model, the *Markov source* or *memory source,* which at the production of a new event can still remember which event it has previously generated.

This kind of source is described inadequately by only the probable occurrence of its events; the modeling here occurs with consideration of the transfer probabilities. With this in mind, consider the hypothetical source in Fig. 3.8. The conditions which are characteristic of the last previously transmitted event (black or white) are given in the circles. The arrows indicate the transfer probability. The probability that a white pixel follows a black pixel is, in this instance, 0.1; the probability that a black pixel follows a black pixel is 0.9. The absolute probabilities of occurrence of black and white, are—because of the symmetry of the source—both 0.5; black and white occur on average with equal regularity. This seems to present a hopeless case for data compression: If we allow an information content of H for this source, we arrive exactly at $H = 1$ bit/pixel.

We could therefore also assume that the coding of an extension of this source would not be advantageous as it could be coded without extension already by 1 bit/pixel (e.g., 0 = black, 1 = white). But with extension of a Markov source, the probability of occurrence of the

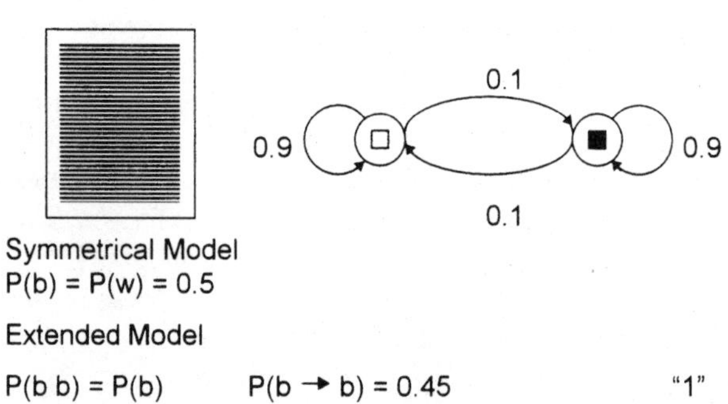

Extended Model

P(b b) = P(b)	P(b → b) = 0.45		"1"
P(w w) = P(w)	P(w → w) = 0.45	Code →	"00"
P(w b) = P(w)	P(w → b) = 0.05		"010"
P(b w) = P(b)	P(b → w) = 0.05		"011"

Figure 3.8 Coding of a Markov source.

pixel pair is calculated slightly differently: the probability of the combination black and white is the product of the probabilities for black (0.5) and after black comes white (0.1). In this way, the probability P and the Huffman code can be calculated once again.

$$P(w, w) = 0.45 \rightarrow c_1 = 1 \quad l_1 = 1$$

$$P(b, b) = 0.45 \rightarrow c_2 = 01 \quad l_2 = 2$$

$$P(b, w) = 0.05 \rightarrow c_3 = 001 \quad l_3 = 3$$

$$P(w, b) = 0.05 \rightarrow c_4 = 000 \quad l_4 = 3$$

Hence, we arrive at the average code-word length

$$L = 1.65 \text{ bit/(2 pixels)} = 0.83 \text{ bit/pixel} < H$$

where $H = 1$ bit. Theoretically this is not possible! The average code-word length is suddenly smaller than the entropy, and we must at this point conclude that the current definition of entropy, in the case of the memory source, is inadequate by being too pessimistic. Nevertheless, there is a more general definition for the entropy H_m (m indicates memory) that also gives a reliable but still achievable lower limit for the average code-word length:

$$Hm = \sum_{\text{All conditions}} P(\text{condition } i) \times H\,(\text{condition } i)$$

For sources containing memory, $H_m < H$ applies.

In the previous example, the two states black and white both have the probability 0.5. The entropy $H(\text{state } i)$ depends only on the transfer probabilities of the single state i (in both cases 0.1 and 0.9). Thus

$$H_m = 0.5 \times H(0.1, 0.9) \times 0.5 \times H(0.1, 0.9) = 0.47 \text{ bit/pixel}$$

which is smaller than L of the current example as theory requires. In a sense, then, memoryless sources have only one state. Thus $H_m = H$ applies to them. For the sake of clarity, we will next summarize the most important results for the memoryless and memory-affected sources.

3.2.3.1 Memoryless sources. The probability that the event $s(n) = s_j$ occurs is independent of which event $s(n-1) = s_i$ has previously occurred:

$$P(s_i \rightarrow s_j) \neq P(s_j)$$

By knowing the event probabilities, we can calculate a lower limit H for the average code-word length L, without having developed a code. If we derive a code with the Huffman algorithm, the following applies for the average code-word length L:

$$H \leq L \leq H + 1$$

If we derive a code for the n-fold extension of this source, the following applies for the average code-word length L for the coding of n pixels:

$$H' = (n \times H) \leq L \leq (n \times H) + 1 = H' + 1$$

This gives per pixel

$$H \leq L \leq H + 1/n$$

3.2.3.2 Sources with memory. The probability of an event occurring depends on which event was previously observed: $P(s_i \rightarrow s_j) \neq P(s_j)$. As shown in Fig. 3.7, $P(b \rightarrow w) = 0.1$ but $P(w) = 0.5$. If we code this source without extension, we do not draw any knowledge from the memory, and the result for the memoryless source applies:

$$H \leq L \leq H + 1$$

If we code this source with extension, the lower, but achievable, limit is given by Hm with $Hm < H$, which results in $H_m < L$. The coding of a memory source without extension is, therefore, even less effective than the extensionless coding of a memoryless source, where a maximum of 1 bit/pixel is lost.

Unfortunately, the extension of sources has an undesirable side effect: if a picture with 256 gray levels is considered as a source, 256 code words are required. If we code the extension by factor n, we use 256^n code words. For the simplest case $n = 2$, we have 65,536 code words, which requires an enormous memory capacity. Considering natural pictures, we discover that—especially in regard to surfaces—similarities do not only exist with directly neighboring pixels. To utilize the memory for natural pictures effectively, the simultaneous coding of only two pixels is far from adequate. Here, we quickly reach the limits of what is achievable.

Nevertheless, it is now known which properties the input signal of the code-word assignment must have in order to generate as few bits as possible and to be simple to encode:

- It must have the smallest possible entropy H, a prominent probability distribution, and the smallest possible amount of discrete amplitudes.
- If possible, it must be memoryless so that the extension can be discarded. The calculation unit (Fig. 3.1) must generate a memoryless

signal $g(n)$ from the memory-affected signal $s(n)$ with the same information content. The coding of $g(n)$ without extension is then as effective as a coding with extension of $s(n)$. The task of the calculation in Fig. 3.1 is now clear: a signal $g(n)$ with exactly these two properties must be derived from the picture signal $s(n)$. The method of making such a calculation is demonstrated later in this chapter.

3.3 Digital Picture Coding: Decorrelation and Quantization

An important goal of picture coding is to represent the picture material with as few bits as possible. The previous sections demonstrated how the bits to be transmitted are generated from a code-word table which gives each gray level a different code word of variable length. To find a code-word assignment (such as a Huffman code) that produces as few bits as possible, the input signal of the code-word table must contain a correspondingly small amount of information. This information content (entropy) can only be drawn from the statistical properties of the source and is independent of how the source is going to be encoded.

As stated in the first Shannon theorem, it is, in principle, possible to find a code-word assignment for every source that generates a bit stream with an average data rate that can get close to but cannot be smaller than the information content of the source. What is important is that the correct source model is chosen for the design of the code.

Theoretically at least, that would solve the problem of the optimum lossless coding. However, as commonly occurs in practice, a problem arises: in natural pictures, the influence of neighboring pixels—what can be called the memory—extends to affect a large number of pixels. To utilize this memory, the gray levels of neighboring pixels must be coded with a common code word, which can be called an *extension*. But an appropriate extension is, as has been shown previously, technically impossible to realize, because the required number of code words is too large. The input signal $g(n)$ of the code-word table should therefore preferably no longer have memory, as opposed to the picture signal $s(n)$. An important task of the calculation unit which has to be followed by the code-word assignment is thereby defined:

- The equation must produce a signal $g(n)$ from the memory-affected picture signal $s(n)$, which is without memory but has the same information content as the frame signal $s(n)$. When that occurs,

$g(n)$ can be encoded without extension almost as effectively as $s(n)$ with extension (which practically would be impossible).

This elimination of memory is possible for certain source models and is referred to as *decorrelation*. It is still a reversible operation and is explained in the following chapter. Quite often, however, decorrelation is not enough to achieve the required data rate. In this case the decorrelated signal is quantized before the code-word assignment takes place. The quantization is the irreversible part of the operation. A loss of information occurs that is explained further in the following sections.

3.3.1 Decorrelation: The source loses its memory

Before the memory of a source can be eliminated in a reversible way, it must first be known what the memory looks like. But it is an almost unsolvable task to determine the structure of an arbitrary memory source (Markov source). There is, however, a simplified source model, the so-called *linear process*, whose memory can be accessed when the source signal is analyzed. Although this model is not perfect in describing the memory of the picture signal, it is satisfactory to work with.

A linear process can be imagined as the linear filtering of a signal containing no memory (Fig. 3.9). We simply assume that the picture signal $s(n)$ of the source to be encoded Q_2 is derived through the linear filtering process of the unobservable signal $z(n)$ without memory, which comes from the source Q_1. This model is not so far from reality as it appears. Through low-pass filtering, for example, such a noise signal $z(n)$, in which no relation between successive pixels exists, can be smoothed out. This results in the creation of a similar neighboring pixel and thus a memory. As this filtering is a reversible process,

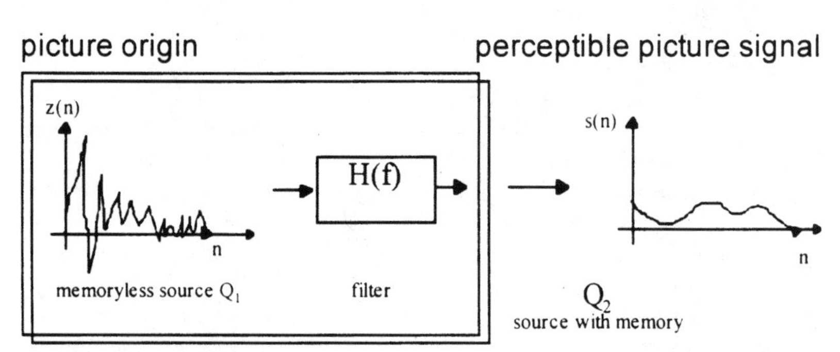

Figure 3.9 Simplified model of memory association of a picture source as a linear process.

there can be no information loss or gain. Therefore Q_1 and Q_2 have the same information content.

Because Q_2 is a memory source, the effective coding of this source is rather difficult due to the previously named basic problems. It would have been simpler to encode Q_1 because this could be encoded without much loss in efficiency and also without extension. But unfortunately Q_1 cannot be observed so easily. Now the calculation unit is applied for the first time. A possible solution to this problem is to find the inverse filter $H^{-1}(f)$ of the model's filter $H(f)$, through analysis of the picture signal $s(n)$. This filter will generate an output signal, ideally one without memory, from the memory-affected input signal $s(n)$, as illustrated in Fig. 3.10. Because a signal without memory has a constant power density over its spectral range (*white noise*), the design problem is to find a filter that smooths out the spectrum of $s(n)$ as much as possible.

If a suitable inverse filter $H^{-1}(f)$ is found, $s(n)$ can be filtered before encoding and the output signal without memory can then be encoded without extension. The signal $s(n)$ is reconstructed at the decoder by filtering $z(n)$, this time with $H(f)$. The widely used coding technique *differential pulse-code modulation* (DPCM) builds on this consideration and will be explained through example.

3.3.2 Predictive coding

Figure 3.11 shows a possible realization of the construction: a two-dimensional digitized picture with 256 gray levels is read out in lines to obtain the sequential signal $s(n)$. The decorrelation filter $H^{-1}(f)$ is arranged in a way that a prediction—a predicted value—is subtracted from the signal value $s(n)$. The resulting value of $z(n)$ of this subtraction is referred to as the *prediction error*. This prediction for $s(n)$ comprises, in the simplest case, only the previous signal value $s(n-1)$, but it can also be composed of a linear combination of several previous

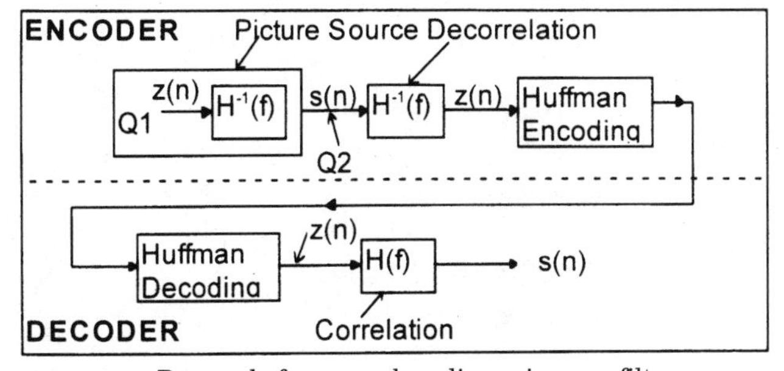

Figure 3.10 Removal of memory by a linear inverse filter.

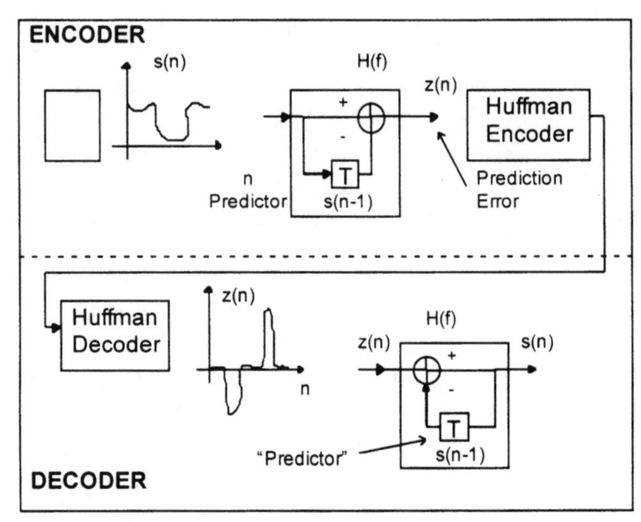

Figure 3.11 Example implementation of lossless DPCM.

signal values (e.g., 0.5×left-hand neighbor + 0.5×neighbor above). The calculation of a linear combination is nothing more than a digital filtering process. We speak, in fact, of a *prediction filter,* which takes the place of the simple delay T in Fig. 3.11. It can be shown that in addition to its characteristic of making the spectrum as white as possible, the optimum prediction filter also minimizes the signal power of the prediction error.

That a simple delay can give good results as a prediction filter becomes clear when we consider that natural pictures are composed of plain surfaces and contours. The differences of neighboring pixels within these surfaces are small, so that $z(n)$ has small signal values. Only contours such as edges and corners generate large differences. Therefore $z(n)$ contains a prominent distribution density (often small amplitudes, seldom large amplitudes) and obviously a small information content. Because of this decorrelation, the entropy H as defined previously has diminished from 7 to about 3 bits for typical picture contents, which results in benefits for the Huffman coding in terms of efficiency.

The variant of DPCM as described here is also called lossless DPCM, because as long as the codec works errorfree, the input and output signals of the codec are exactly equal. A suggestion for such a system with its own code-word table can be found in Ref. 8. The further variants of DPCM in its loss-prone form, as well as with prediction in the time dimension with movement compensation, will be described later.

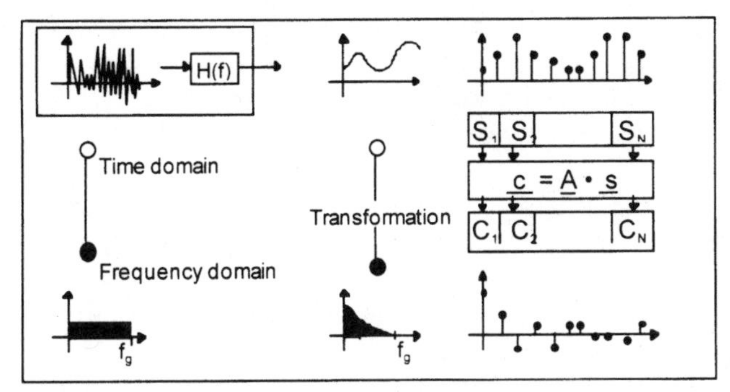

Figure 3.12 Transformation—another way to remove the memory.

3.3.3 Basic idea of transformation coding

A further method of utilizing the memory of the signal to be encoded is to transform the signal by blocks with a suitable transformation matrix **A** into the frequency domain, as illustrated in Fig. 3.12. Like filtering, the transformation is a fully reversible operation: through the transformation of a block of N neighboring signal values $\mathbf{s} = [s(n), s(n-1), \ldots, s(n-N+1)]$, a block of exactly N coefficients $\mathbf{c} = (c1, c2, \ldots, cN)$ is produced, which are later converted into the N signal values again by means of the inverse transformation. The list of values **s** and **c** can be regarded each as an N-dimensional vector. Transformation, then, is nothing more than the multiplication of this data vector **s** with a matrix **A**.

Transformation:

$$\boldsymbol{c} = \boldsymbol{A} \times \boldsymbol{s}$$

Inverse transformation:

$$\boldsymbol{s} = \boldsymbol{A}^{-1} \times \boldsymbol{c}$$

For coding, the coefficients have the advantage over pixels in that they are mutually decorrelated in proximity fashion, and thus have no memory. It is known that the low local frequencies (for instance in surfaces) are predominant, and therefore the main part of the picture information is concentrated only on a few low-frequency spectral components. That is why decorrelation goes hand in hand with a concentration of power density on few coefficients. All coefficients $c_1, c_2, \ldots, c_N$ have, therefore, different probability distributions in which each coefficient needs its own Huffman code. An example of a transformation coding without losses can be found in Ref. 9. Because of its importance, transformation coding will be discussed in more detail in a subsequent chapter.

Several decorrelation methods, such as prediction or transformation, are sometimes cascaded in coding concepts. But it should be clear that a further decorrelation of signals which are more or less decorrelated already may only increase encoding efficiency by a small amount.

Lossless coding demands a technique suitable for applications in which even the smallest coding error cannot be tolerated, such as the archiving of x rays. Such an example presents a much different situation than the transmission of video signals intended for communication or entertainment purposes. Further, transmission channels typically have a constant bit rate, while a lossless scheme would want to send more data when there is more picture detail. So, unfortunately, with lossless encoding, we quickly come up against the limits of possibilities. Seldom are compression factors of more than 2 achieved. The reasons for this are:

- Data compression without loss is, of course, only possible if redundancy exists in the frame content, and the statistical characteristics of the frame content remain reasonably stable. In unlucky cases, a *data expansion* can arise from the data compression operation.
- The number of possible values of the decorrelated signal can be very large with large transformation matrices and prediction filters of a higher order, leading to a high number of necessary code words. For example, in Fig. 3.10, the range of values of $s(n)$ $\{0,\ldots, 255\}$ to $g(n)$ $\{-255,\ldots, 255\}$ has doubled.
- Natural pictures that result from the linear filtering of a source without memory do not always deliver discrete amplitude values only. Therefore, there is no guarantee that the entropy of the decorrelated signal will become smaller with increasing decorrelation. We can however show that with DPCM, for example, the power of the signal to be encoded becomes smaller with increasing decorrelation—which is especially interesting with respect to quantization.
- Irrelevancies in the pictures hide a large potential for data compression that lossless coding does not utilize.

Here, the irreversible part of the encoding process helps further; the decorrelated signal—the prediction error or the coefficients—are quantized before their coding.

3.3.4 Quantization of decorrelated signals

Figure 3.13 demonstrates the principle of quantization. The signal amplitude is mapped to a *reconstruction level* (e.g., 4 on the y axis)

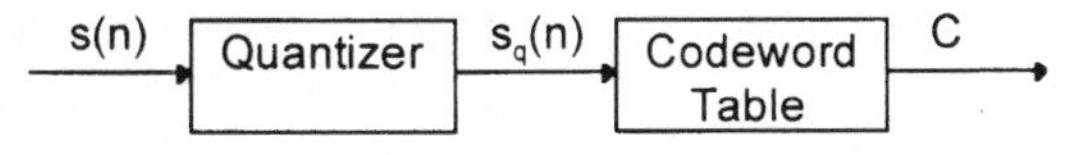

Example:

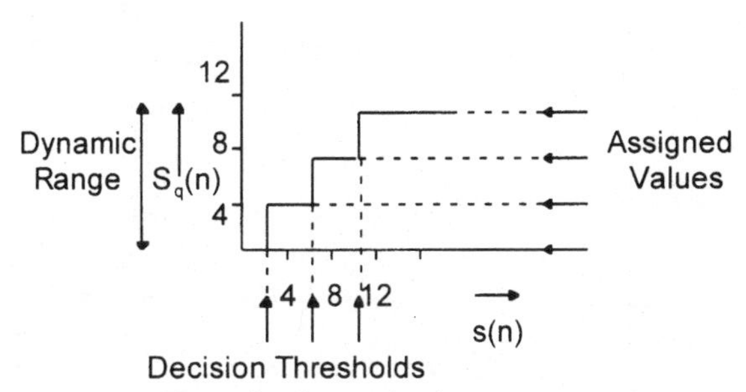

Figure 3.13 Example of quantization.

representing the range between two decision levels (e.g., 2 and 6 on the x axis). The values of the reconstruction levels are then encoded with fixed or variable word lengths, depending on the approach. The receiver replaces the true (but unknown) signal value $s(n)$ with the allocated reconstruction levels $sq(n)$ in the decoded signal. The resulting error $q(n) = s(n) - sq(n)$ is called the *quantization error.* The distance between the smallest and the largest reconstruction level is the *dynamic range* of the quantizer. A simple form of quantization is the familiar rounding off to whole numbers. Quantizers for codes with variable word lengths will be henceforth called *entropy-coded quantizers* (ECQ), and those for codes with fixed word lengths will be called *fixed-length-coded quantizers* (FLCQ).

Before dealing with the design of quantizers it is helpful to become familiar with the notion of variance and signal power. The variance σ^2 of a picture signal $s(n)$ is an important characteristic and shows how significantly the signal varies around its mean value by giving its mean square deviation. Under the assumption that a signal having large amplitude variations has to be encoded with a high number of bits, this value gives an idea of the information content of the signal. In contrast to the idea of variance is the notion of signal power P, which is the mean squared amplitude of the signal. Signal power P is identical to σ^2 in cases in which the mean value of the picture is zero. Because in principle it is not relevant how the gray scale or color scale of the picture content is mapped to the corresponding steps, it is possible to define a mapping where the mean value of the signal is always zero. For example a range of gray may be coded from -127 (black) to $+127$ (white). This allows the common use of the term *signal power* instead of *variance,* which is typical in the literature. As a reminder, the mathematical definitions of the mean value *mu*, the signal power *P* and the variance σ^2 are

$$mux = \frac{1}{N} \sum_{n=1}^{N} x(n)$$

$$P = \frac{1}{N} \sum_{n=1}^{N} x(n) \exp 2$$

$$\sigma^2 = \frac{1}{(N-1)} \sum_{n=1}^{N} [(x(n)-mux)]^2$$

3.3.5 Step-by-step design of the quantizer

Step 1. State an *error criterion.* The error criterion could be the entity which damages the subjective picture quality. A simpler but also less explanatory criterion is the well-known SNR (signal-to-noise ratio), the logarithmic relationship between the signal power of $s(n)$ σ^2 (s) and the mean square quantization error, that is, the signal power of $q(n)$, σ^2 (q):

$$\text{SNR} = 10 \times \log \frac{\sigma^2\,(s)}{\sigma^2\,(q)}$$

Quantization errors can have diverse effects on the picture content. The results of direct quantization of gray levels, transformation coefficients, or the use of DPCM may each lead to different effects in the picture.

Step 2. State the maximum tolerable error, given the error criteria (for example, the maximum allowable SNR or maximum subjective quality loss).

Step 3. Design a quantizer that produces the smallest data rate, taking into consideration the permissible error. Design parameters include the numbers and positions of the reconstruction levels, as well as the position of the decision levels between the reconstruction levels. The determination of these parameters depends on a subjective selection of which of the masking effects inherent to the human visual system is the most suitable for use.

The execution of step 3 depends on whether the quantized signal is to be encoded with a variable (ECQ) or fixed word length (FLCQ). In the case of ECQ, finding the smallest data rate depends on finding the quantizer which, within the specified range of error, minimizes the entropy. In the case of FLCQ, the bit rate depends directly on the number of levels of the quantizer. For example, for 16 steps, 4 bits are

S(n)	$S_q(n)$	Code: Constant Word Length	Code: Variable Word Length
...2	0	"00"	"1"
2...6	4	"01"	"01"
6...10	8	"10"	"001"
10...	12	"11"	"000"

Figure 3.14 Code assignment of representing values. Constant word length is simpler while variable word length is more efficient.

required. By specifying the error tolerance, the number of levels would thereby be minimized. In both cases it would also be possible to state the data rate instead of defining the tolerable error. This leads to the requirement that the design of the quantizer minimize the error at a certain data rate.

Coding with a fixed word length simplifies code-word assignment, the decoding of code words, and error protection (Fig. 3.14.) Additionally, during transmission with a fixed bit rate, there is no need for a data buffer because the data rate is constant and is not dependent on the picture content. These advantages should not disguise the fact that coding with a fixed word length is not an optimum solution. This is because of the fact that to minimize the mean data rate, the calculation unit (now with the given error) must minimize the entropy of the signal to be encoded, which will then be encoded with variable word length. In practice, ECQ thus comes more to the front. Basically, we discriminate between linear and nonlinear quantizers, as discussed next.

3.3.6 Nonlinear quantizers

Figure 3.15 shows the characteristic of a nonlinear quantizer. As we can see, this sort of quantizer is characterized by the different sizes that the quantization steps can have. As shown in Fig. 3.14, it can be realized with a table of values but also through two nonlinear transfer functions with a linear quantizer switched in between. This topic will arise again in our discussion of sound as the key element of the well-known audio *compander* (*com*press + ex*pand*).

How do we design or optimize a nonlinear quantizer? An approach from a psychovisual viewpoint follows. Typically, we require that the coding error falls below the perception threshold and minimizes under this condition the number of reconstruction levels (FLCQ) or the entropy (ECQ). With DPCM, for example, highly contrasting pic-

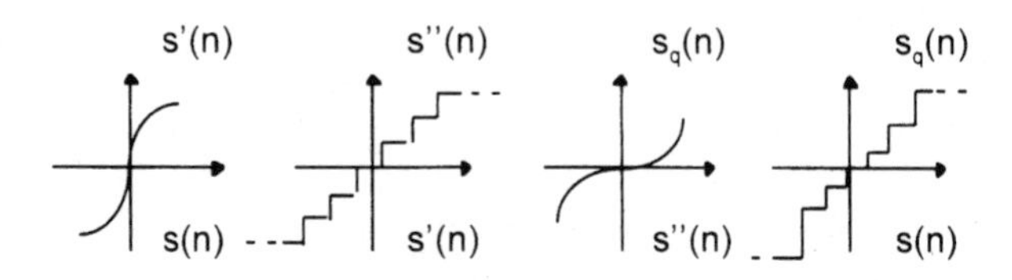

Figure 3.15 Nonlinear quantization. Implementation options: companding and linear quantization.

ture areas and contours generate a large prediction error, and it is precisely such picture areas that are suitable to disguise coding errors. Furthermore, large amplitudes of the prediction error can be quantized more coarsely. We get similar quantizers with DPCM for ECQ and FLCQ, with around 20 to 30 reconstruction levels.[10] Higher horizontal, vertical, and all diagonal spatial frequencies are quantized coarsely with transformation coding, as they are not so well-perceived.

3.3.7 Max-Quantizers, Lloyd algorithm

Other less laborious design techniques are based on a statistical approach. Here, the attempt is made to adjust the characteristic of the quantizers of the signal for a given number of levels (FLCQ), so that the $\sigma^2(q)$ (mean square error) becomes minimal. These quantizers are named after their discoverer.[13] They have the interesting property that the quantized signal value receives exactly the same reconstruction level as the next successive value, because minimizing the error is all that matters. By stating the reconstruction levels, these quantizers are now fully described. The *Lloyd algorithm*[11] also builds on this simple idea to optimize such quantizers. A representative number of samples of the signals to be quantized (e.g., prediction errors) are then generated. Thereafter the reconstruction levels are initialized, for example, with equidistant values. The algorithm then proceeds as follows:

- *Step 1.* Each reconstruction level is given to the sample that lies next to it. (The samples are not quantized.)
- *Step 2.* The new reconstruction levels are derived from the arithmetic mean values of the samples that are given to them. The new reconstruction level, with respect to these samples, thus generates a smaller error. The algorithm in step 1 is continued with the new reconstruction levels.

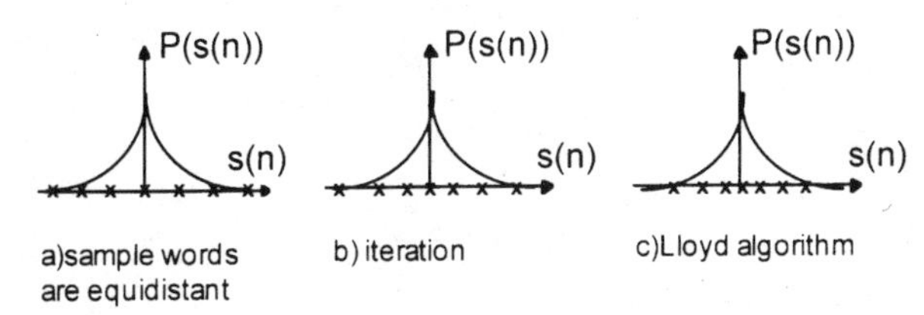

Figure 3.16 Iterative optimization with the Lloyd algorithm.

These two iteration steps change places until the reconstruction levels have converged sufficiently. Consequently, we get many levels in such ranges which are strongly represented in the input signal, such as small signal values in the prediction signal. This procedure also produces small quantization errors more frequently than large errors, so, on average, smaller quantization errors are produced (Fig. 3.16).

Although the Lloyd algorithm was developed to optimize FLCQs, we can improve the efficiency of the quantizer through Huffman coding. A simpler ECQ would be desirable in this case. There is also an algorithm for optimizing ECQs, that is, for minimizing $\sigma^2(q)$ at a given entropy of the quantizer output.[12] With the ECQ, quantization might not assign a sample to the next adjacent reconstruction level if there is a similarly good level that can be coded with significantly fewer bits. Therefore, the decision levels must also be defined to fully describe the ECQ.

The optimization of the quantizer using subjective criteria is in any case to be preferred because the statistical method only functions if the statistical characteristics of the signal remain stable. A quantizer that is optimum for a certain signal statistic can be extremely bad for another statistic. Additionally, the minimizing of $\sigma^2(q)$ leads to undesirable errors due to a dynamic range which is too small. This applies above all for the FLCQ, because it must cope with as few as possible reconstruction levels. These errors are on average small but very visible.

3.3.8 Linear quantizers

Contrary to the nonlinear quantizer, the linear quantizer, shown in Fig. 3.17, has equidistant reconstruction levels and thus does not leave much room for optimization. Nevertheless, it is frequently applied because of the following reasons:

- It is simple to realize, and to vary we only need to do a rounding-off operation and to give a multiplier to the coder and decoder, respectively. Figure 3.17 shows the quantization of the value 5.6 with a step size of $w = 2$.

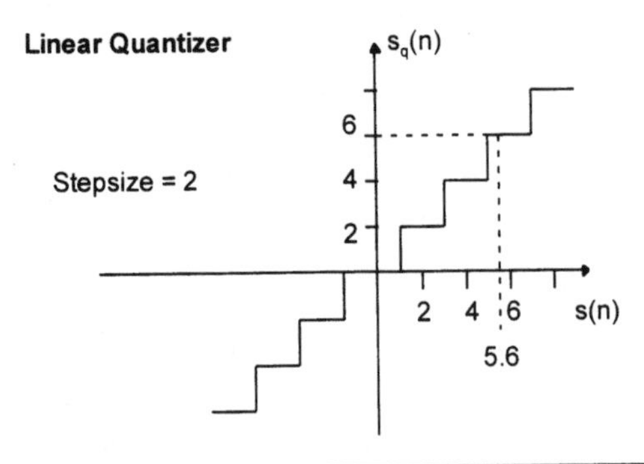

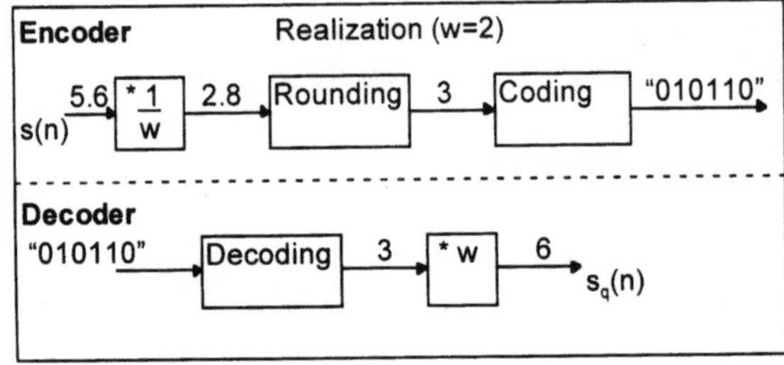

Figure 3.17 A linear quantizer.

- With the mean square error (MSE) as the error criterion, a linear ECQ is superior to a nonlinear FLCQ. However, when encoded with a fixed word length, the linear quantizer is inefficient if the distribution density of the input signal is great, as for example in Fig. 3.16.
- In any case, with the statistical optimization of nonlinear quantizers, they converge with an increasing number of reconstruction levels into a linear quantizer. The equidistant quantization steps are then optimum for a higher data rate.
- It is often difficult to devise a nonlinear quantizer from a visual aspect. There are, therefore, no reliable guidelines over whether and how the perception of a quantization error in transformation coefficients depends on the amplitude of the coefficients. Further, changes in the signal statistics must be reckoned with. In these cases, the linear quantizer is the neutral solution.

3.3.9 Why decorrelate before quantization?

There is one remaining question in this introduction to quantization: Why should we decorrelate when the decorrelated signal is already quantized before coding? It has been shown that a calculation of the approximate number of bits necessary for coding can be made before a code-word assignment is devised. But how is it possible to anticipate before quantization the number of bits we need to encode the signal after quantization? The so-called *rate distortion theory* has been busy with this problem for some time. The type of simple relations found with errorfree coding are, unfortunately, not available here, especially not for subjective error criteria. Nevertheless, we can give an approximate rule-of-thumb formula for the simple error criterion MSE:[13]

$$R(\sigma_q^2) \approx \frac{1}{6}\, 10 \log\left(\frac{\sigma_s^2}{\sigma_q^2}\right) + c = \frac{1}{6}\,\mathrm{SNR} + c$$

Variable $R(\sigma_q^2)$ is the data rate required to encode the quantized signal, if an MSE of σ_q^2 is to be achieved during quantization. We could also write $H(\sigma_q^2)$ *for an ECQ. The constant* c lies approximately between -0.5 and 1.2 and is dependent on the distribution density of the input signal and the choice of quantization concept: a small c is obtained for a peak distribution density if a variable word length is used for encoding. It should be noted that a data rate increase of 1 bit means an SNR improvement of approximately 6 dB.

Despite this formula only having an approximate validity, it helps explain the effectiveness of a decorrelation through calculation of difference signals or through transformation. For example, with DPCM the signal power of the quantized signal is reduced through decorrelation. Additionally, the distribution density of the difference signal is much more prominent (leading to a small c) than the distribution density of the picture signal itself. We need a smaller data rate when coding the decorrelated signal, for the same $^2(q)$. Finally, we must ensure that through the inverse calculation unit $H(f)$ (Fig. 3.10) only the power of the signal is increased, and not the power of the error; otherwise nothing will have been gained. Looking at Fig. 3.18, assuming the predictions $\hat{s}_1(n)$ and $\hat{s}_2(n)$ in the coder and decoder are identical, the decoded output signal is the sum of the input signal $s(n)$ and the quantization error $q(n)$; the mean signal power will be increased through the addition of the predictor but not that of the error. Unfortunately, there is a small snag to this: when we look more closely, we will see that at the receiver, the quantization error—as part of the decoded signal—is fed back into the predictor. As a result of this, the predictions of the coder and the decoder diverge and the decoded signal contains an additional error because of the incorrect prediction. The following chapter deals with DPCM techniques and how this problem can be dealt with in a simple manner.

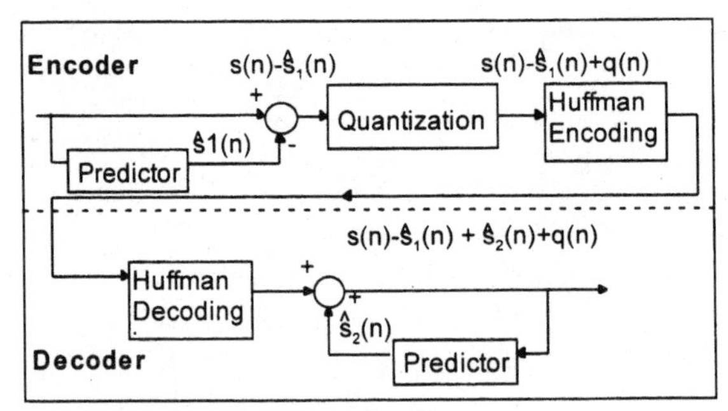

Figure 3.18 Quantization and coding.

3.4 References

1. Shannon, C. E., "A Mathematical Theory of Communication," *Bell Syst. Tech. J.,* vol. 27, 1948.
2. Hamming, R. W., *Coding and Information Theory,* Prentice Hall, New York, 1980.
3. Gallager, R. G., "Variations on a Theme by Huffman," *IEEE Trans. Inform. Theory,* vol. 24, November 1978, pp. 668–674.
4. Ziv, J., "Coding Theorems for Individual Sequences," *IEEE Trans. Inform. Theory,* vol. 24, July 1978, pp. 405–412. (s.a. Welch, T. A., "A Technique for High Performance Data Compression," *IEEE Computer,* vol. 17, June 1984, pp. 8–19.)
5. Rissanen, J., and G. G. Langdon, "Arithmetic Coding," *IBM J. Res. Develop.,* vol. 23, March 1979, pp. 149–162. (s.a. Mitchell, J. L., and W. B. Pennebaker, "Software Implementation of the Q-Coder," *IBM J. Res. Develop.,* vol. 32, November 1988, pp. 753– 774.)
6. Netravali, A. N., and B. G. Haskell, *Digital Pictures, Representation and Compression,* Plenum Press, 1988.
7. Jayant, N. S., and P. Noll, *Digital Coding of Waveforms,* Prentice Hall, New York, 1984.
8. ISO/IEC JTC1/SC2/WG8: JPEG Technical Specification, Revision 5, January 1990, pp. 68–74.
9. Shah I. A., O. A. Akiwumi-Assami, and B. Johnson, "A Chipset for Lossless Image Compression," *IEEE J. Solid State Circuits,* vol. 26, no. 3, March 1991, pp. 237–244.
10. Netravali, A. N., and B. G. Haskell, *Digital Pictures, Representation and Coding,* Plenum Press, 1988.
11. Fleischer, P. E., "Sufficient Conditions for Achieving Minimum Distortion in a Quantizer," *IEEE Int. Convention Rec.,* part 1, 1964, pp. 104–111.
12. Chou, P. A., T. Lookabaugh, and R. M. Gray, "Entropy-Constrained Vector Quantization," *IEEE Trans. on ASSP,* vol. 37, January 1989, pp. 31–42.
13. Jayant, N. S., and P. Noll, *Digital Coding of Waveforms,* Prentice Hall, New York, 1984.

Chapter

4

Video Compression Techniques: Differential Pulse-Code Modulation

Differential pulse-code modulation (DPCM) is an important and well-established method of data compression. In its basic form, it achieves reasonable compression results without much effort. Further, most of the methods of data compression with large compression factors have a DPCM loop [Moving Pictures Experts Group (MPEG), for example] at the core. In principle, DPCM transfers only the difference (D) between the actual input signal and its best predicted value.

4.1 DPCM as a Simple Method of Decorrelation

In the previous chapter, we briefly introduced DPCM, an encoding process that decorrelates the memory-affected picture signal via a filter. The benefit of the decorrelation process is that the signal receives an amplitude distribution that is concentrated around zero. A bit-saving transmission with entropy coding can only be achieved through this amplitude distribution. For decorrelation, only a special type of filter can be used, in which the output signal is derived from the difference between the input signal and a so-called prediction signal. This has the advantage that the predictor P can be intuitively interpreted: the predictor must give a prediction for the next pixel to be transmitted from the observation of the previously transmitted pixels.

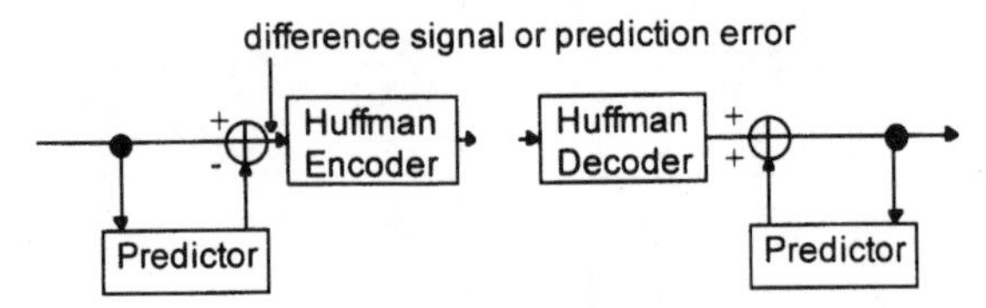

Figure 4.1 Block diagram of a lossless DPCM system.

The better these predictions are, the smaller is the mean difference between the true input signal and its prediction that has to be transmitted, and therefore also the lower the bit rate. This difference is carried by the information which is still unknown to the receiver. The difference signal is often referred to as *prediction error* because it corresponds exactly to the deviation between the prediction and the actual picture signal to be transmitted. (The prediction error should not be confused with the quantization error.) If we complete the system with the receiver, which also derives its prediction from the transmitted and earlier decoded pixels, and then corrects it with the prediction error received, we have a scheme of the lossless DPCM (Fig. 4.1).

4.1.1 Error-affected DPCM

Because of different signals having a typical entropy of about 3 bits with a simple DPCM, we cannot transmit all frame information with a compression factor of over 2. However, we can create errors as long as they are not perceived by the human visual system. This is the principle of the loss-affected DPCM: the entropy of the prediction error will be considerably further reduced by the quantization of the prediction error before transmission. Because a smaller number of possible different events is to be transmitted, the entropy—and thus the data rate—falls. Unfortunately, so does the picture quality. Finally, through quantization, the prediction error is superimposed with an undesirable but unavoidable quantization error.

The following problem is even more serious than the issue of picture quality. During the time that the predictor in the encoder has the input signal at its disposal, each pixel at the output of the decoder is compiled from the combination of three factors: (1) the prediction of the decoder predictor, (2) the prediction error, and (3) the added quantization error. As a result, each pixel is modified to some extent. This results in different signals being received at the transmitter side and the receiver side. Despite this, the modified output signal of the decoder manages to make a useful prediction in the receiver, but it is no longer exactly the same as that in the transmitter. The next pixel will thus be derived quite differently, and so on. This divergence results in the predictors drifting apart.

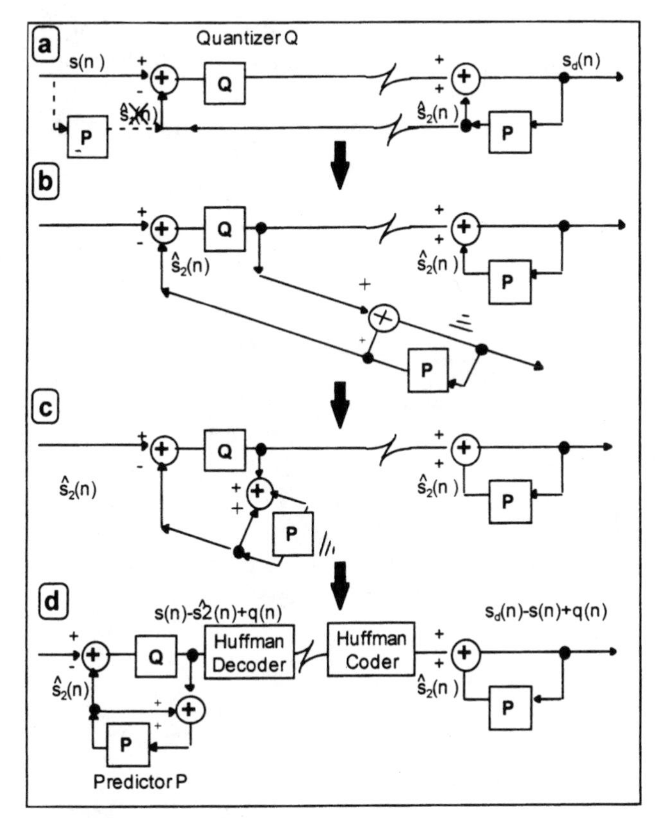

Figure 4.2 Error-affected DPCM: (*a*) prediction utilization of the receiver, (*b*) reproduction of the receiver, (*c*) integration into the transmitter, (*d*) correct, loss-affected DPCM.

It would be better to know the prediction $\hat{s}_2(n)$ of the receiver at the transmitter already and apply it there instead of its own prediction $\hat{s}_1(n)$ (Fig. 4.2). In this case, instead of returning the prediction signal, we duplicate it in the receiver and send it back to the transmitter, thereby deriving the typical structure of the error-affected DPCM transmission.

It is clear from this construction that the predictors of both the transmitter and the receiver cannot drift apart. Both predictors are supplied with exactly the same input signal, as no error occurs on the way. In the hopefully rare case of a transmission error, only the receiver predictor will be affected.

4.1.2 What do we get from the prediction?

The interesting question arises: How many bits do we save by the error-containing DPCM transmission, when the picture must have the same quality as a directly quantized and transmitted picture? We can make a rough estimate of this so-called prediction gain with the rule of thumb introduced previously, in which the bit rate can be roughly estimated depending on the relationship between the signal and the quantization error power resulting from a quantization. If we want to allow a mean square error $\sigma^2(q)$ by direct quantization of a picture, in order to code the quantized signal, we need a data rate of R_{PCM}

$$R_{\text{PCM}} = \tfrac{1}{6} \times 10 \times \log \frac{\sigma_{\text{PCM}}{}^2}{\sigma^2(q)} + c_1$$

The variable $\sigma_{\text{PCM}}{}^2$ is the power of the DC-adjusted (mean value = 0) picture signal. The constant c_1 depends on the quantization and the distribution density. Similarly, we can also estimate the required bit rate for the DPCM. Admittedly, we use the much smaller power of the prediction error $\sigma_{\text{DPCM}}{}^2$. For a fair comparison of the bit rates, the same MSE, namely $\sigma^2(q)$, is allowed in the decoded frame. The quantization error $q(n)$ of the prediction error gives the same error in the decoded picture (for the decoded picture signal see Fig. 4.2d):

$$s_d(n) = s(n) - \hat{S}_2(n) + q(n) + \hat{S}_2(n) = s(n) + q(n)$$

Therefore, it applies that for the data rate of the DPCM

$$R_{\text{DPCM}} = \tfrac{1}{6} \times 10 \times \log \frac{\sigma_{\text{PCM}}{}^2}{\sigma^2(q)} + c_2$$

Although with Huffman coding of the prediction error, the constant c_2 is in general slightly smaller than c_1, we can approximate c_2 by setting it equal to c_1 and derive the prediction gain:

$$R_{\text{PCM}} - R_{\text{DPCM}} = \tfrac{1}{6} \times 10 \times \log \frac{\sigma_{\text{PCM}}{}^2}{\sigma_{\text{DPCM}}{}^2}$$

As was to be expected, the smaller the power of the prediction error, the more advantageous is the DPCM compared to direct quantization and transmission.

4.1.3 Predictors for television pictures

Predictors for the data compression of television pictures can be arrived at with a sound mind. The result can be optically controlled by viewing the signal after the generation of the difference on a monitor: the darker the so-called prediction-error picture is, the better the predictor. During the design, we try to build as much knowledge as possible about the source characteristics of the television picture into the predictor but also to apply as much information as possible from the environment of the pixels to be transmitted.

But only those pixels which are already known (all pixels to the left of the momentary pixel, as well as those in all lines above it) can be made use of by the receiver. As a reminder, we previously explained that in present television standards the frame consists of two fields, with each field containing every other line of the frame. The first field contains all odd lines corresponding to a certain point in time, and the second field contains all even lines corresponding to a point in time just between the previous field (with odd lines only) and the next field (also with odd lines only). For a reliable prediction in this case, both the geometric relations have to be taken into account and the information from the correct point in time has to be used (and not mixed up with the pictorial information of other fields). Because current video systems are interlaced, we must basically discriminate between those which work within one half of the frame (*intrafield predictor*, Fig. 4.3*a*) and those which, with the help of a field memory, compile the full frame and from that arrive at a prediction (*intraframe prediction,* Fig. 4.3*b*).

Intrafield predictors cannot make use of all vertical neighboring relationships, due to the lack of access to the line immediately above. We see, however, that with an intraframe predictor the similarity with vertical neighboring pixels is lost with small horizontal movement, because succeeding picture lines come from different phases of

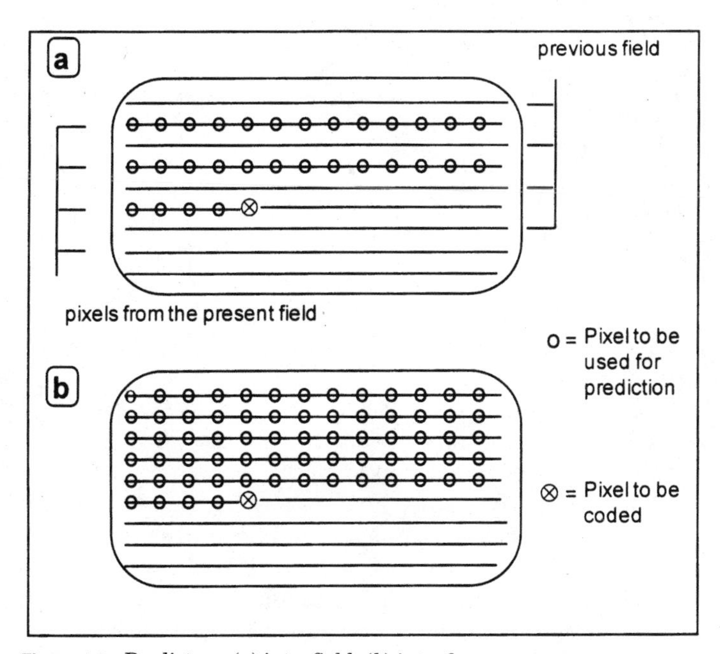

Figure 4.3 Predictors: (*a*) intrafield, (*b*) intraframe.

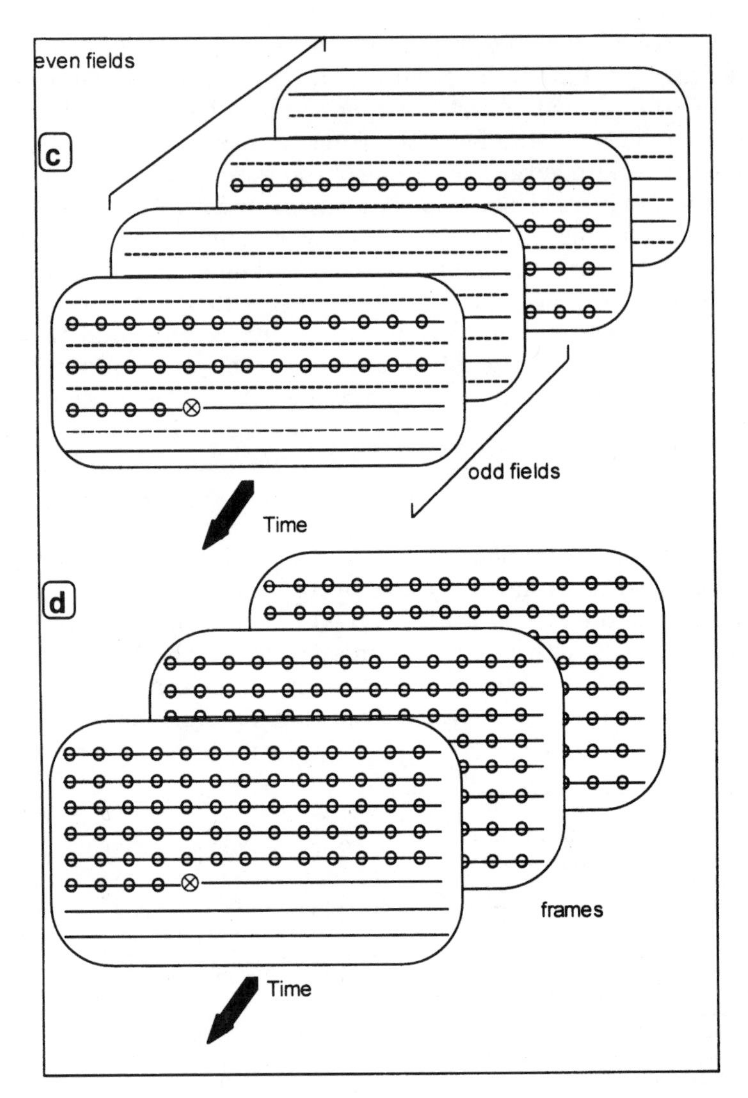

Figure 4.3 (*Cont.*) (*c*) interfield, (*d*) interframe.

movement. (In this case, small teeth appear on vertical structures in the fully compiled frame.) In this situation, intraframe predictors may generate a high power of prediction error.

Beyond the momentary full frame, further underlying frames can be recaptured. If only frames from the same field are used, we speak of an *interfield* predictor. If any parts of both fields of each frame are to be used, we speak of an *interframe* predictor. It should be noted that with frames lying behind, lines also below the momentary pixel of the frame are accessible because they are already known by the receiver (Fig. 4.3*c* and *d*). Such interpredictors, however, require

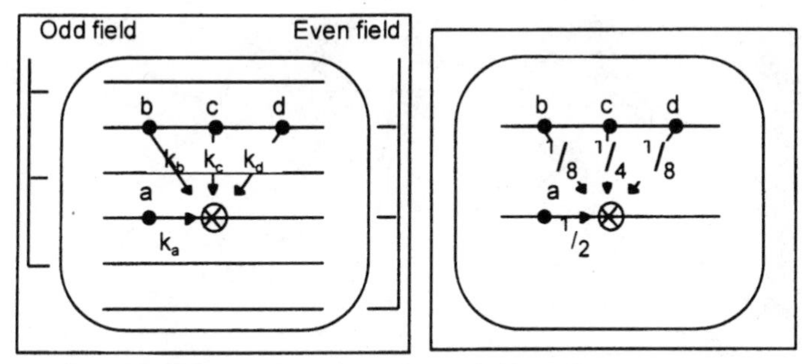

Figure 4.4 Simple two-dimensional intrafield predictor.

extra mechanisms such as movement compensation to target smaller prediction errors.

4.1.4 Intraprediction

Predictors that have been realized to date use only a small part of all pixels admissible for the prediction, as can be seen in an intrafield predictor shown in Fig. 4.4. The prediction value $\hat{s}_2(n)$ is compiled from the sum of pixels *a, b, c,* and *d,* weighted with k_a, k_b, k_c, and k_d, of the re-decoded picture at the transmitter. Mostly, $k_a+k_b+k_c+k_d = 1$ is used because the pixel to be transmitted is, on average, neither smaller nor larger than its surroundings.

A purely horizontally oriented predictor (*one-dimensional* (1D) predictor, with $k_a = 1$, $k_b = k_c = k_d = 0$) assumes that horizontally adjacent pixels are equal. Such a predictor can give good results with horizontal lines and surfaces but not with vertical edges, as shown in Fig. 4.5.

The *two-dimensional* (2D) predictor suffices in those cases where the pixel above the momentary line is also used. The attempt is to optimize the coefficients $k_a, \ldots, k_d$ of 2D predictors by statistical methods with respect to small prediction errors. The nearer the given pixel is to the picture spot, the more heavily it is weighted. The pixel on its immediate left has the strongest influence. Because this prediction more or less represents the mean value of the immediate surroundings, it will give good results in surfaces and, with limitations, also in horizontal edges. Conversely, in vertical edges it gives poor results as can be seen in the prediction error frame of this predictor in Fig. 4.5*c*. An optimum situation for upright edges would be a vertical 1D predictor with $k_c = 1$, $k_a = k_b = k_d = 0$. An even better result could be realized through the use of an adaptation mechanism that recognizes which picture structure should be transferred and selects the most suitable predictor accordingly.

(a)

(b)

Figure 4.5 Predictor results: (a) original picture, (b) 1D predictor.

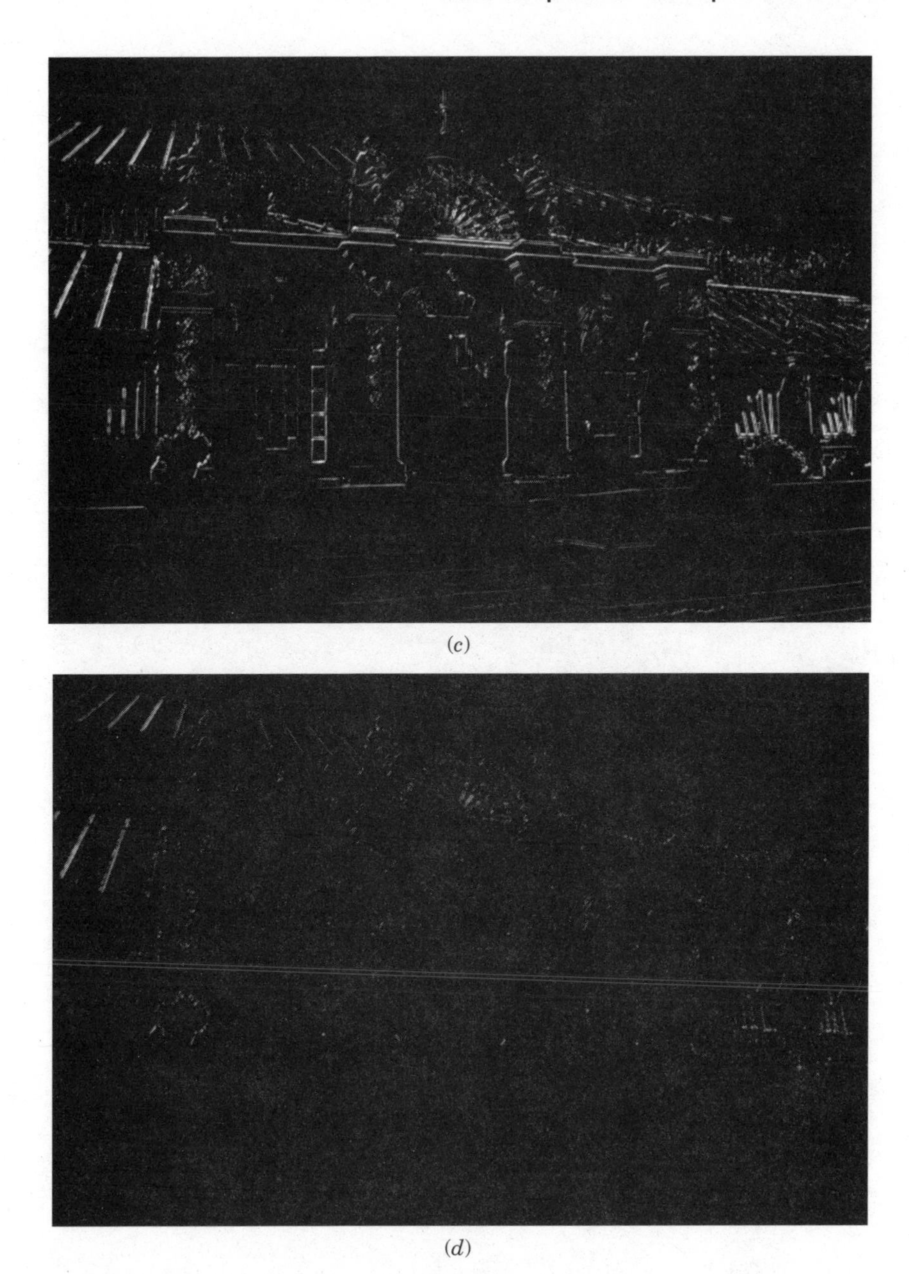

(*c*)

(*d*)

Figure 4.5 (*Cont.*) (*c*) 2D predictor, (*d*) adaptive predictor.

4.1.5 Adaptive predictors

For television pictures, the model concept that the picture signal is switched between several generation processes, e.g., the processes' horizontal edges and vertical edges, is applicable. Better predictors pick up the factual bill: depending on the local picture object, switching occurs between different predictors. Because the predictor adjusts to the picture signal, we speak of *adaptive* predictors. For the horizontal edge predictor, we can use for example the horizontal 1D predictor. In vertical edges, we switch to the vertical 1D predictor. It can be seen in Fig. 4.5*d* that such a predictor generates a dark prediction-error frame. To also generate a good prediction with diagonal structures, we need a diagonal predictor, and so on. In practice we restrict ourselves to a maximum of four different predictors (horizontal edges, vertical edges, surface, and texture). For the selection mechanism, two factors must be distinguished:

- *Causal predictors.* These select only those pixels for the adaptation that may also be used for the prediction, which are already known to the receiver. The receiver recognizes, synchronous with the transmitter, the right predictor from the picture data already received. In the simple case of only two predictors, a comparison will be made to see if changes in gray level occur between pixels *a* and *b* that are as great as those between *c* and *d*. If that is the case, the predictor for a horizontal edge will be selected. More detailed methods carry out a recognition of the picture structures and, thus, can switch more confidently between a greater number of predictors.
- Noncausal predictors. In this case, the pixels that are selected for the predictor are as yet unknown to the receiver. The output signal of the predictor is quasi-influenced by information from the future. This information must be given to the receiver in advance via an additional side channel, with its respective data rate, in order to guarantee continued synchronization between transmitter and receiver. To ensure that the necessary data rate is not too high, the frame is divided into large blocks in which the same predictor is used.

4.1.6 Transmission errors of the channel

If the channel generates transmission errors with the DPCM, typical picture disturbances occur depending on the type of predictor in use. If only one single gray level in the prediction error frame is decoded incorrectly, e.g., one horizontal 1D predictor in the receiver gives wrong predictions for the remainder of the line, the predictors drift apart. This error will appear as horizontal stripes (Fig. 4.6). In a similar fashion, the 2D predictor distributes throughout its message a diagonal pattern that extends to the picture border, as shown in Fig. 4.7. The parts of the picture that are disturbed due to transmission errors can be made smaller by reducing the sum of the coefficients of the predictor to a value less than 1, so that the error in the receiver predictor loop fades away, in a process known as *leakage.* The transmitter and receiver predictors must, however, be identical so that the sum of the coefficients of the transmitter predictor is less than 1. The transmitter predictor underestimates (on average) the signal to be transmitted with the result that the performance of the prediction error and the bit rate increase.

With a transmission error in the DPCM system with adaptive predictors, the receiver selects (in the event of error) perhaps even the wrong predictor branch, so that all following pixels are wrongly interpreted. Under certain circumstances the rest of the frame is destroyed. Most DPCM systems work with an entropy coding of the prediction error, so that the code words have different lengths. Single-bit errors then lead to the code words not being properly separated and recognizable and to larger picture areas becoming unusable. We must therefore introduce at predetermined locations (in addition to channel error-protection and error-correction mechanisms), clearly

Figure 4.6 The effects of transmission errors with a 1D predictor.

Figure 4.7 The effects of transmission errors with a 2D predictor.

identifiable synchronization words, with which the receiver can recognize the start of a new code-word sequence.

4.1.7 Quantizers for the prediction error

In transmitting the luminance signal of normal pictures, the signal value is directly proportional to brightness. Thus there is no range of values in which errors cause disturbances that are less disturbing subjectively. (Because of gamma precompensation, the perception threshold for errors with direct quantization in dark and light areas is at least similar.) For the quantization of prediction errors, we can better exploit the perception characteristics of the eye. Prediction errors only contain nominal amplitudes on edges and, in general, in detailed parts of a picture. In surfaces, there are almost no differences. The limited perceptive qualities of the eye for coding errors in such parts of the frame (using the Mach effect and masking) need to be exploited for a coarser quantization of the edges and the detailed areas of the frame. To start the discussion, the effect of quantization errors with DPCM will be investigated by using a simple DPCM system.

The simple DPCM system works with a 1D predictor and, for the time being, with a linear quantizer. The quantizer only has a few steps to ensure that it generates a low bit rate. If a large dynamic range is required, the result will be a coarser quantization (Fig. 4.8*a*). With such a quantizer, a homogeneous surface with small difference signals will be transmitted, corresponding with the left-hand part of Fig. 4.9. Because the quantizer reacts to small changes with its full step height, the output signal is actually too large. With the following pixel, an attempt will be made to correct the quantization error by transmitting a value that is too small and vice versa. The jumping around of the quantizer output signal appears in the decoded picture as noise on a homogeneous surface, a *granular noise,* where it is especially visible (Fig. 4.10). If we decide to use a finer quantization in the design, and the resulting dynamic range of the quantizer is too small (Fig. 4.8*b*), errors will be generated during transmission of the edges in the right-hand part of Fig. 4.9. The jump in brightness leads to large difference signals at the edges, and the quantizer becomes overloaded. Because the gray levels in the decoded picture (from pixel to pixel) cannot change more than the maximum quantizer reconstruction level, the slope of the decoded edge lags behind the real-edge slope. This effect is known as *slope overload,* and is shown in Fig. 4.11. As the lines are processed independently from each other, the slope overload can occur at different places on a line basis. Vertical edges will be torn. During transmission of a moving picture, this behavior results in a busy-edge effect.

On the other hand, a nonlinear quantizer (Fig. 4.9*c*) can mostly avoid these problems. The small output signals in homogeneous surfaces will

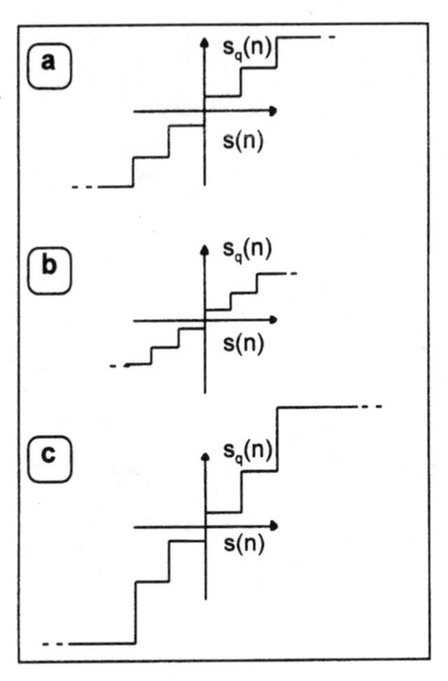

Figure 4.8 Quantizing step considerations: (*a*) large quantizing steps can create granular noise; (*b*) smaller quantizing steps can suffer from slope overload; (*c*) a nonlinear quantizer provides small steps for minor value changes and large steps for major transitions.

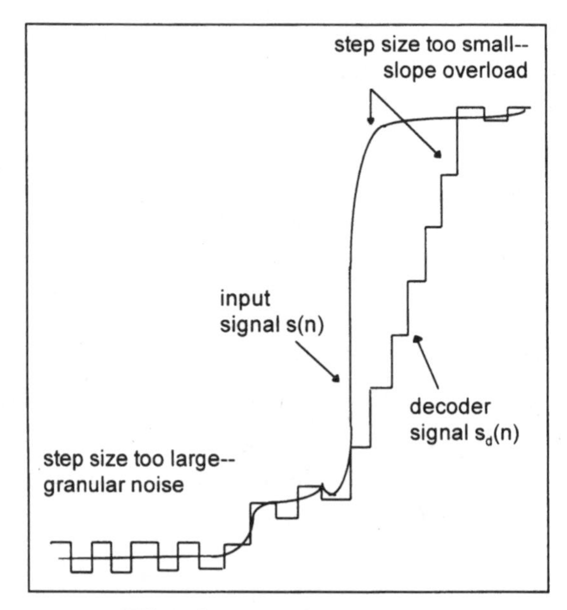

Figure 4.9 Effect of quantization.

Figure 4.10 Granular noise.

Figure 4.11 Slope overload.

be quantized finely enough so that the noise stays under the visible limits. There is nevertheless still a large maximum signal available for edges. The extremely coarse quantization of large amplitudes on edges is not noticeable because of the Mach effect, which effectively masks quantizing errors in the edge regions. But how do we derive the decision levels and reconstruction levels of this quantizer?

The Lloyd algorithm was introduced earlier as a possible means of designing a nonlinear quantizer with a given number of steps (FLCQ). By using a measured amplitude distribution of the prediction-error picture, this method generates a nonlinear quantizer for which the resulting mean square quantization error will be minimized at the output. Because the amplitudes that occur in the prediction-error picture are almost all low, small amplitudes are also generated for the MSE. A quantizer is, therefore, realized that quantizes small amplitudes very finely. Almost all the decision thresholds fall in the region of small amplitudes. Although this quantizer leads to a small MSE, it quite often has a dynamic range that is too limited, leading to slope overload, even with small edges. Considered solely on the basis of subjective picture quality, the MSE is not very descriptive. If the MSE is weighted according to the size of the prediction-error signal, the perceptive characteristics of the eye can be taken into account.

By carrying out perception tests in which the visibility thresholds for given coding errors of typical edge profiles are measured, we can make better use of the masking characteristics for errors. We can determine the prediction error that occurs for typical picture material (e.g., through simulation) and set the quantization threshold so that the error always lies below the measured visibility limit. It is nevertheless a safe and good method to design the quantizer intuitively and to optimize it by perception tests on various frame material.

Table 4.1 contains reconstruction levels and decision levels of a quantizer for DPCM that was optimized by psychooptical tests. Based on the Huffman coding for the reconstruction levels, the entropy of the quantized signal is about 2.6 bits/pixel.[1] In this example only the positive reconstruction levels and decision levels are given because the quantizer is symmetrical.

4.1.8 Adaptive quantizers

For the prediction, the optimum predictor was selected depending on the picture content to be transmitted. Adaptivity is also an advantage with respect to the quantizer: the perception threshold for coding errors is higher in heavily detailed areas of the frame. We can save bits by switching to a coarser quantizer with fewer steps in those regions. A measure for the amount of detail can be obtained from the gray level distribution of the preceding pixels around the pixel to be transmitted. The larger the maximum absolute gray level difference of pixel pairs that can be generated from all the pixels a, b, c, and d in Fig. 4.4, the coarser will be the quantization. In practice, we limit

TABLE 4.1 Reconstruction Levels (Upper Line), Decision Levels, and Range Between (Lower Line) for DPCM

Reconstruction levels	Range between reconstruction and decision levels
0	0–2
4	3–6
9	7–11
14	12–16
19	17–21
24	22–27
31	28–34
36	35–41
45	42–48
53	49–57
62	58–66
71	67–75
80	76–84
91	85–255

ourselves mostly to four different quantizers.[1] As with predictors, we can distinguish between causal adaptivity and noncausal adaptivity (with a suitable side channel).

Depending on the quantizer in use, the input signal of the Huffman code indicates differences in probability distribution of the reconstruction levels. In actual fact, for coding purposes, a different Huffman table should be used for each possible predictor-quantizer combination. However, practice has shown that this is not necessary as the deviations are sufficiently small.[1]

Whereas we can expect compression factors up to only about 2 from the lossless DPCM (depending on the picture material), factors of up to 3.5 are achievable with the loss-affected intra-DPCM, as long as the coding errors are required to be invisible at a normal viewing distance. For larger compression factors, there is however a better predictor: the last frame sent. Admittedly, small movements generate difference pictures with very high performance, so that for the interframe-DPCM, a key element is missing—motion compensation—which is the subject of the next section.

4.2 Interframe Coding of Picture Sequences and Motion Compensation

In natural scenes, successive frames of a sequence are very similar. We can take advantage of this fact during data compression and save bits by transmitting only the small differences that appear from frame to frame. When movement occurs, the similarity is reduced and the data rate increases. But the picture content only appears to have changed: the same objects continue to appear in successive frames, only they appear in another place. Is it possible to reproduce the similarities between the frames whereby the moved objects are put right again? The following describes how, by means of motion-compensated interframe DPCM further improvement can be made in compression for the digital transmission of television pictures.

Essential for motion compensation is the measurement of motion. This can be realized, for instance, with the frequently used *block-matching* technique. In addition to block matching, other methods of motion measurement will be briefly introduced. In addition to picture data compression, motion-controlled signal processing has proven valuable in other applications, including standard conversion, field-rate up-conversion, tape-to-film transfer (and vice versa), and noise reduction. The requirements of these techniques with respect to motion measurement will be shown to be substantially different from those of data compression.

4.2.1 Predictors for the DPCM

The previous section dealt with intraframe DPCM. All variants of DPCM have in common the fact that the picture itself is no longer transmitted, but rather the so-called prediction error is transmitted. This prediction error is the smallest difference between the momentary pixel to be transmitted and a best possible prediction. The prediction is derived from some previously transmitted pixels. The better the predictions are, the better the compression result will be. We have already shown predictors that exclusively evaluate the momentary frame. An even better prediction is obtained for stationary picture content in a sequence of frames if the pixel having the same x and y coordinates as in the previous frame is used for prediction. Such a predictor belongs to the class known as *interframe predictors.* The interframe predictors we will now examine generate the prediction only from the previous frame.

Figure 4.12 shows how the interframe predictor works: the previous frame is stored in a digital frame memory for the duration of a full frame period. The previous frame can thus be subtracted from the momentary frame to be transmitted. As both frames are very similar, the resulting difference picture hardly changes in the regions where no motion occurs and can be transmitted with a significant saving of bits. The previous frame is already known to the receiver. Because the receiver also has a frame memory, it can predict the momentary frame by addition of the difference frame received.

4.2.2 Why motion compensation?

In the scene, a ball has moved from the left in the previous frame to the right in the momentary frame. If we consider the content of the same pixel in successive frames, we will see that the regions with motion in the momentary and the previous frame come from parts of

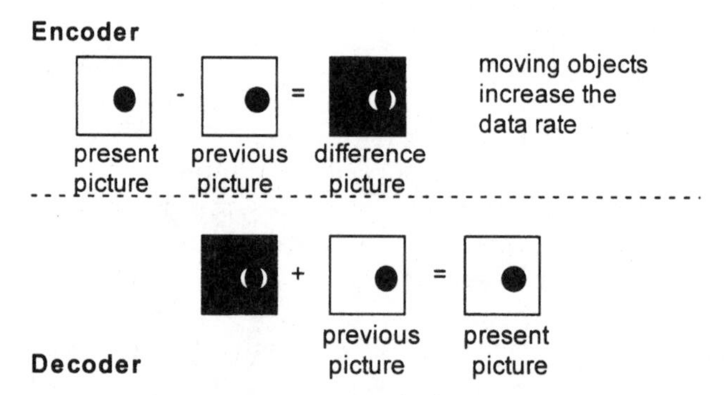

Figure 4.12 Interframe DPCM without motion compensation.

the scene that are remote from each other. There are hardly any similarities between these pixels, so the difference picture signal receives a significant boost in the regions which are in motion. As the signal power increases, so too does the data rate. As a single camera moves across a scene with strongly detailed picture content, the result is that more data must be transmitted for the difference frame than for the original frame. To make use of the underlying frame as a predictor, the previous and momentary frames must be matched by means of the so-called motion compensation. The coder takes the object from the old frame, as if it were from a set of building blocks, and moves it to the correct position.

4.2.3 Interframe DPCM with motion compensation

Motion measurement is a subcomponent of motion compensation, in which the motion of all objects from the old to the new frame are measured and the borders of the objects defined. The ball that moves from left to right can be identified in Fig. 4.13. After measurement, the ball is shifted in the previous frame in relation to its motion, and a motion-compensated frame is obtained. As we can imagine, the result

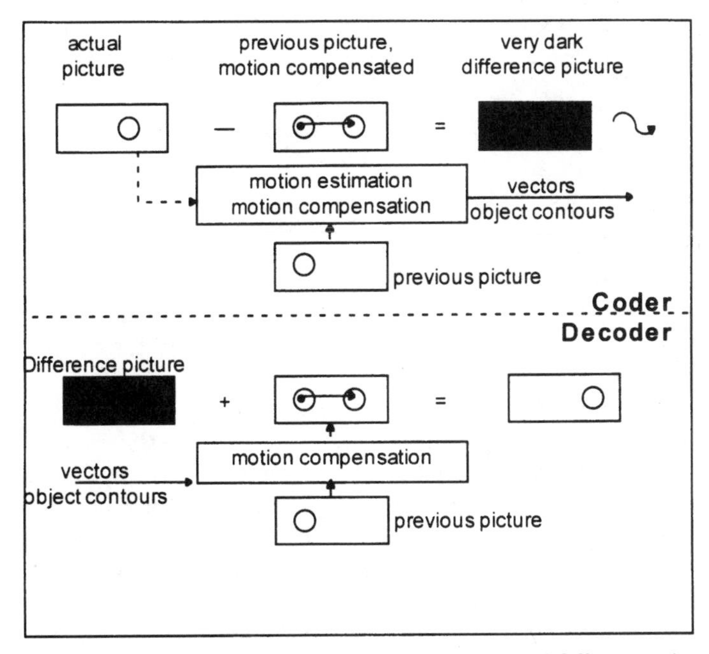

Figure 4.13 Motion compensation reduces the data of difference pictures when there is motion.

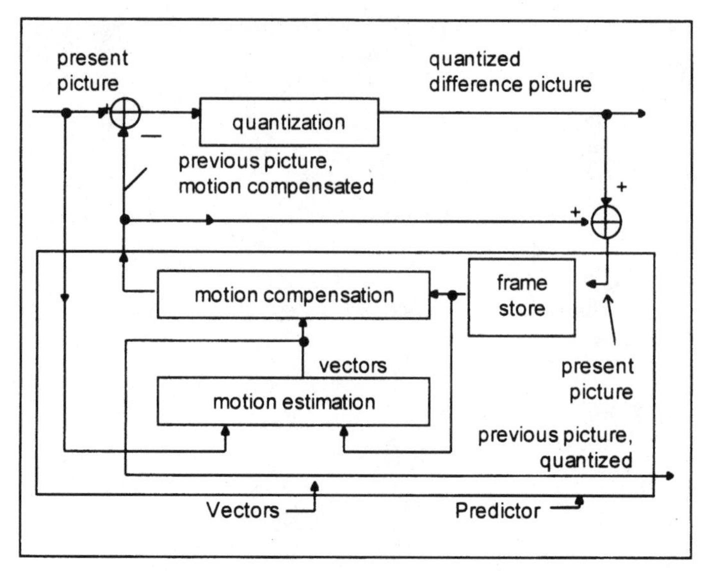

Figure 4.14 Motion-compensated interframe DPCM encoder.

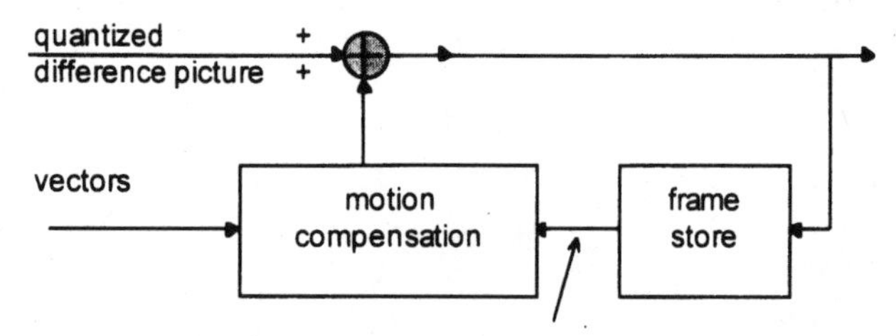

Figure 4.15 Motion-compensated interframe DPCM decoder.

of this motion compensation is a very dark difference picture, even in the regions that are in motion. The entire signal flow can be traced via Fig. 4.14, in which the total interframe-DPCM system for motion compensation is shown.

The predictor (shown in the large box) in the coder comprises, in addition to a frame memory, a block which carries out the motion measurement by comparing the old and new frames. Following this, a motion-compensated frame is generated at the output of the predictor with the help of the motion information (object outlines and motion vectors). Notice that the motion measurement made at the transmitter uses the momentary frame that is not yet known to the receiver. The motion measurement cannot, therefore, take place simultaneously at the receiver, despite it being needed for the receiver prediction (Fig. 4.15). The motion information has to be transmitted additionally via a side channel. The motion-compensated interframe DPCM works, therefore, with a noncausal predictor.

4.2.4 Methods of estimating motion via block matching

A motion compensator that totally neutralizes all motion would produce such good prediction frames that practically no power would remain in the difference picture. We need a relatively large amount of data for a detailed description of motion but only a relatively small amount for the difference frame. Admittedly, even using state-of-the-art techniques it is not yet possible to identify and measure the motion of any object from general frame sources. We have to settle for simplified picture models, as for example with the frequently used block-matching technique. Despite suboptimal motion compensation, the data rate necessary for the difference picture is much less than without motion compensation. Furthermore, we have the advantage of a particularly simple and therefore bit-saving motion description. This partly makes up for the signal power of the difference picture, which is not completely minimized.

4.2.5 A motion estimator using block matching

In data compression applications, the block-matching motion estimator has at its disposal each new frame to be transmitted from directly adjacent objects that all have the same dimensions. Additionally, the objects can only move uniformly in a single direction in the two-dimensional plane. The frame to be transmitted is, therefore, divided into a number of rectangular pattern blocks (Fig. 4.16), which are processed successively. The motion predictor assumes that the pattern blocks can only be moved by a maximum amount in the x and y directions. For each pattern block there is a search region in which the pattern block can be found in the previous frame, due to the underlying model. Using a constant step width, the pattern block is shifted through successive positions within the search region, and each position is compared with the old picture.

The shift in position, also called *displacement,* that achieves the best similarity or matching is interpreted as the sought-after motion. The block of the motion-compensated frame is then filled in with the content of the block belonging to the previous frame, which in turn was the best match under the pattern block being sought previously. In this way, the motion-compensated frame is brought blockwise as near as possible to the momentary frame.

Figure 4.16 Motion compensation with block matching.

The x and y components of the displacement are sent to the receiver via the side channel, so that a motion-compensated frame can also be constructed from the old frame. Carrying out this task on the content of the previous frame, and thus at the known picture, is the fundamental advantage of this coding method.

The data rate for the vectors now depends on the size of the search region, and thus on the maximum shift, as well as the desired accuracy of the vectors. Contours of the objects do not have to be transmitted because all objects are identically sized rectangles.

The *displaced frame difference* (DFD) is used as a measure for comparing the similarity between the displaced pattern block and the underlying previous frame. The pixels of the pattern block are subtracted from the underlying pixels of the previous frame. The absolute values of the differences are then added. The smaller the differences are, the smaller is the DFD and the better the pattern block matches the position in question. The values of the DFD can be arranged with respect to the x and y displacement in a so-called *correlation surface.* The motion of the object is related to the position of the minimum of the DFD. It is called *full-search* block matching when looking for the minimum DFD, and all possible positions of the correlation surface are scanned with a fixed step width and then enumerated.

The underlying model simplifies images to such an extent that the question must be asked: How useful are the predicted motion vectors for coding? Here the specific demands of image data compression distinguish themselves from other applications. For example, a high-quality interframe interpolation is used to ensure that the motion vector of any pixel corresponds to the actual motion of a real object in the picture. A wrong vector leads immediately to incorrectly represented objects in the interpolated frame or field, which can be very disturbing. Conversely, with data compression, an attempt is made to minimize, on average, the power of the difference picture. To do this, the coder uses a block of the old picture that differs as little as possible from the momentary block of the actual picture, and this is exactly what block matching does. We can also interpret block matching in a different way: we search for any block which above all has as little difference to the momentary block as possible. From this point of view, we are not so interested in the true motion at all.

Figure 4.17*a, b,* and *c* gives an impression of how much motion compensation reduces the power of the difference picture through use of block matching. In Fig. 4.17*a,* a train is moving from left to right. Figure 4.17*b* shows the difference picture without motion compensation. Figure 4.17*c* shows that the scene is much darker with activated motion compensation. The blocks on the borders of moving scene elements (the contours of the train, for example) contain both a still

(*a*)

Figure 4.17 Motion compensation results: (*a*) original picture.

(*b*)

(*c*)

Figure 4.17 (*Cont.*) (*b*) difference picture without motion compensation, (*c*) difference picture with motion compensation.

background as well as moving parts of the picture, which shows the limitations of block matching for motion compensation. But again, for this application there is no real interest in the true motion at all; motion estimation is only used as a tool to minimize the prediction error and thus allow for efficient coding.

4.2.6 How much effort is necessary for block matching?

With integrated-circuit memory cost continually decreasing, storing the picture in memory is not too daunting. However, block matching requires an enormous amount of calculating power. For each frame, we must

- Investigate all pattern blocks
- Examine the entire search region for each of these pattern blocks
- Calculate every possible displacement and the DFD in the search region

A brief discussion of the parameters of this method, such as pattern block size, search region, and vector accuracy, follows.

With pattern block size, a compromise has to be reached between the quality of the motion-compensated prediction picture and the data rate of the motion information. The smaller the blocks are, the better the picture model reflects reality, and both the power of the difference picture and the data rate fall. The larger the blocks are, the fewer vectors must be transmitted per frame. Therefore, larger blocks, with dimensions ranging from 8×8 up to 32×32 pixels, are usually used.[8]

The scope of a reasonable search region depends on the maximum speed of motion of the picture scene (for example 1 frame width per second). A search region of ± 16 horizontal and ± 8 vertical pixels is frequently used.

When searching for a minimum of the DFD, the smaller the step width is, the more accurate are the vectors. For example, a step width of one pixel yields an accuracy of ± 0.5 pixel. While some methods may be satisfied with that, a significantly better data compression is not achieved until an accuracy of ± 0.25 pixel is used.[9] A performance comparison is given in Ref. 6. Searching shifts using a step width of half of a pixel produces more accurate vectors, but the word length of the x and y components increases by 1 bit, and therefore the data rate for the motion information is increased. Further, the additional intermediate positions must be interpolated and the search positions checked.

Block matching is composed of subtracting, adding, and giving the absolute values of the results of these operations. The number of necessary operations for block matching for a standard television system amount to

$$\left(45\times 36\ \frac{\text{blocks}}{\text{frame}}\right)\times\left(32\times 16\ \frac{\text{search positions}}{\text{block}}\right)\times\left(16\times 16\ \frac{\text{pixels}}{\text{block}}\right)$$

$$\times\left(3\ \frac{\text{operations}}{\text{pixel}}\right)\times\left(25\ \frac{\text{frames}}{\text{s}}\right) = 15 \text{ billion } \frac{\text{operations}}{\text{s}}$$

To reduce this gigantic number, full-search block matching is extended with intelligence which reduces the number of search positions without allowing the motion vectors to become too inaccurate.

4.2.7 Block-matching techniques with reduced calculation requirement

With the so-called quick full search,[6] the search region is investigated around the zero-vector with full accuracy, so that good motion compensation with reduced data rate is achieved for slow-moving parts of the picture. For motion at a higher speed, searching takes place with double step width, that is, with less accuracy. The decreased motion compensation leads to a higher power of the difference picture in the quickly moving picture blocks. As stationary regions are much more frequent in natural scenes, the higher data rate caused by the comparatively fewer quickly moving objects is not so important. This does not apply, however, when, for example, a camera is panned across a scene. Nevertheless, the data rate can also be held constant here through the use of a coarser quantization of moving blocks but at the expense of picture quality. This can be tolerated under certain circumstances, as long as the eye is not able to follow the movement—which is the case with small monitor screens. Errors in moving objects are not perceived as well because of the inability of the eye to resolve them.

Other methods assume that the values of the DFD rise almost always from a global minimum in all directions (Fig. 4.18). This is why we usually find the global minimum when we move into the direction of the decreasing DFDs. With the *2D logarithmic search,* several DFDs at a larger distance are calculated using the zero vector as a starting point. Only the sample by which the smallest DFD was obtained will be pursued. If it turns out that a sample from outside the borders provides the minimum DFD, the unchanged sampling distance will be tried again. If the central sample was the best, further sampling will take place more closely around it. The process lasts until the distances between the samples is so small that they

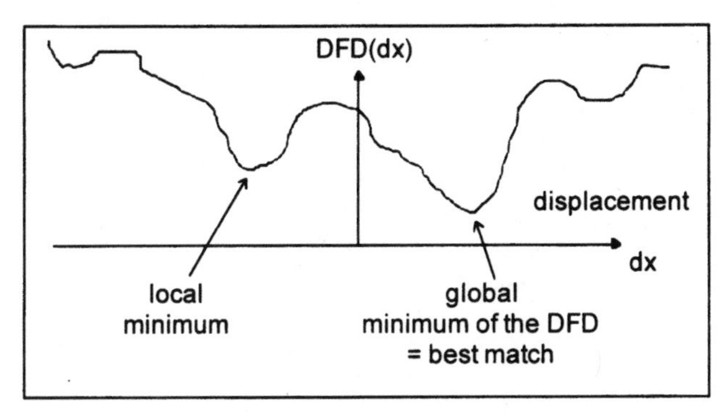

Figure 4.18 Correlation function with one-dimensional block matching.

correspond with the desired vector accuracy. The *three-step method*[4] is a similar process. Unfortunately, these gradient methods frequently end in local minima of the DFD.

Better results are given by the *hierarchical search.*[6] In the initial run, the total search region is covered in search of the proximity of the global minimum of the DFD, with a distance increment of two pixels. Afterward, a finer search matrix is selected and the search continues in the proximity of the presumed minimum until the exact position is found. With this method, local minima rarely lead to faulty predictions. Though intelligent search strategies clearly reduce the amount of calculation work (particularly in computer simulations), control of the search means additional work during integrated-circuit design.[10] Figure 4.19 summarizes the common search strategies.

4.2.8 Other methods of motion estimation

Despite not being used much for data compression, the *phase-correlation* and *differential* methods of motion estimation are useful in other applications.[5] The advantage of phase correlation, compared to full-search block matching, is the reduced calculation requirement with high-speed motion[11] and the far-reaching insensitivity to changes in the illumination of the objects.

The phase-correlation method is based on a convolution of the two pictures under consideration, which can be realized by multiplying the fast Fourier transforms of both pictures after the appropriate amplitude corrections. Figure 4.20 shows a one-dimensional example of how an object displaced from left to right yields a delta impulse after inverse transformation. The distance of this impulse from the zero point is the amount of the displacement. We can find possible motion vectors, therefore, by looking for peaks in the correlation surface. Translatory movements lend themselves well to being detected

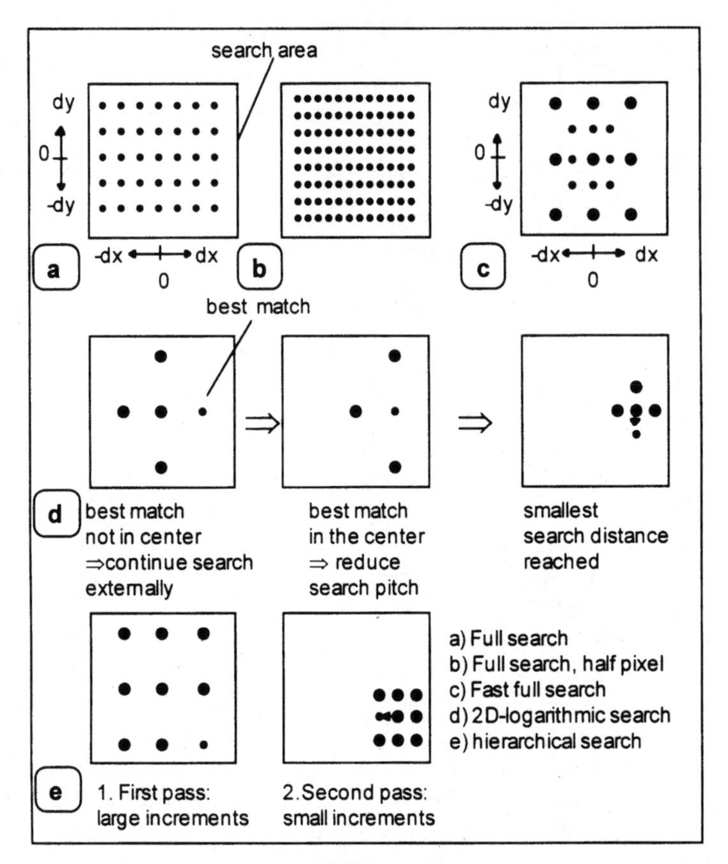

Figure 4.19 Search strategies for different block-matching algorithms.

this way. Admittedly, the peak height reacts sensitively to objects that rotate and/or change their size, as occurs, for example, when a camera zooms in. Therefore, we make use of the phase-correlation method on blocks that are not too large and in which zoom and rotation in general can be approximated through translations. On the other hand, if the blocks are too small, there is a growing influence of artifacts generated by the algorithm itself because the borders of the regions for which the transformation is applied can no longer be neglected. We therefore use a block size of 32 to 128 pixels, horizontally and vertically. Phase correlation supplies a number of possible movements, or *candidate vectors,* for the region being investigated. These originate from the various object movements that occur in the block and also from possible misinterpretations through disturbances. For a further application of the vectors, a test must be made for each pixel, after local vector assignment, to see which vectors are useful.

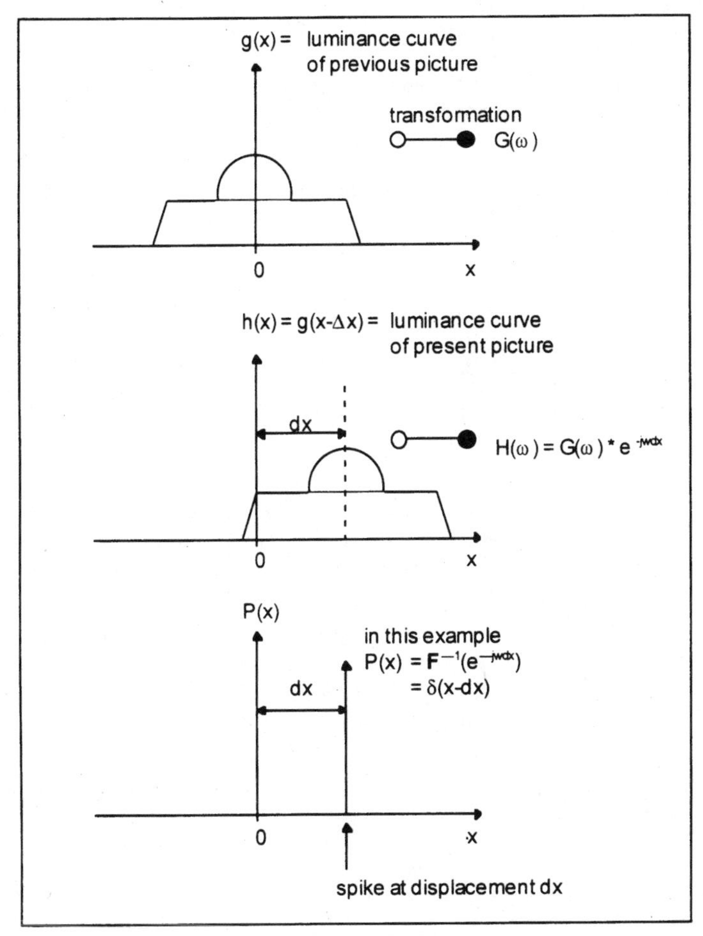

Figure 4.20 One-dimensional example of motion estimation using phase correlation.

The differential methods use the relationship between the gradients of the signal in place and time. The principle is clearly shown in Fig. 4.21 by means of a linear edge gradient of a vertically oriented edge that is displaced purely in the x direction. If the edge moves, the difference between the luminance gradients of the edge in the previous frame $L_0(x)$ and the momentary frame $L_1(x)$ grows in proportion to the displacement of the edge. For a measurement of the movement, the region in which both frames have a linear edge gradient is particularly good for evaluation. In the spots that are used for measurement, we divide the difference in brightness between both frames by the

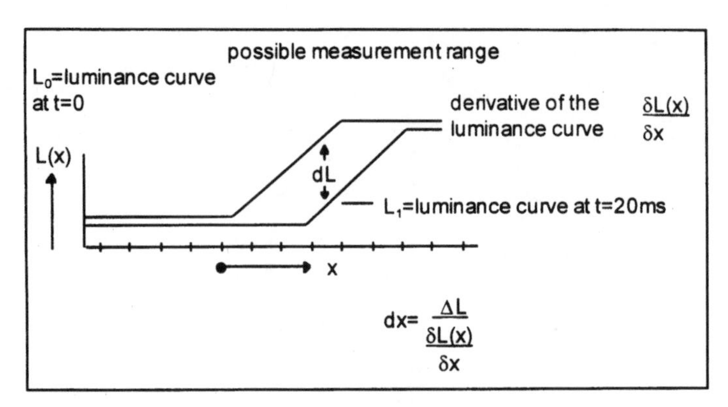

Figure 4.21 Differential motion estimation procedure.

increase in the luminance gradient and then form the average over all the measurement points. As the gradients of the edge must bisect for detection, we can only measure slow movements or else a higher frame sampling rate must be used. The measurement range can be extended by another method that gives a rough prediction of the movement—block matching with a larger step width. Precompensation is carried out with the prediction, so the edges lie on top of each other and a measurement is possible. The differential method in this simple form without precompensation is relatively easy to implement due to the displacement being calculable, and is not—as in block matching—the result of a search requiring much effort.

4.2.9 Disturbing influences on the measurement

The greater the noise affecting television pictures, the more the measured vectors deviate from the actual movements. Detailed picture regions are less affected as they stand out more clearly from the noise.[3] A further source of error is the interlace: if we put together a full frame from two half-frames of different phases of motion, using interframe DPCM in an interlaced system, the resulting motion vectors will possibly be inaccurate. With block matching, the restrictive model requirements make themselves evident: the pattern blocks at the borders of objects can contain different moving parts from the objects and the background, which can lead to incorrect measurements. In addition, the model of exclusively translatory movements, which can be approached via displacements, neglects all other movements such as rotation and zoom. Periodic brightness fluctuations resulting from lighting variations, or much more frequently due to shadows in the picture, also go unnoticed. Periodic structures can provoke errors with almost all methods, because under certain circumstances the wrong periods are identified. These disturbing influences lead to vectors that do not correspond with the true movement. These movement measurement errors are only of secondary importance for the coding, as long as they do not cause the power of the difference picture to rise as outlined earlier. Inaccurate measurements, such as those that arise due to periodic structures, do not necessarily lead to a rise in power, unlike other disturbing influences previously described. With the knowledge and the component cost that we have today, block matching has proven to be the most practical method of motion estimation for data compression.

4.2.10 Motion vectors for general applications

Because of disturbing influences, the methods described previously give only raw vectors, which would be unusable for interframe interpolation. The reliability of the vectors must be increased by plausibility tests of each vector. The local vector assignment must assign the measured vectors to the scanning spots in such a way that the borders of the vector field match up with borders of the object. Although the problem of an obscured background has not been previously discussed, Fig. 4.13 presents such a dilemma: Which frame contents should the motion compensator in the prediction frame select to fill in the region where the ball was in the previous frame? Motion measurement and interframe interpolation demand much more processing power than simple motion compensation for data compression.[7]

Motion compensation makes possible the use of previous frames for prediction with DPCM. Thanks to this, depending on the picture material to be transmitted, a data rate of from 0.5 to 2.8 bits/pixel can be achieved with loss-affected motion-compensated interframe DPCM.[4] This amount of data is still too large for many communication needs, such as terrestrial broadcast of digital television. A further step in this direction is transform coding, which is the subject of the next chapter. But first it must be emphasized that it is the combination of both motion compensation and DPCM that forms the basis of systems which make digital picture communication possible.

4.3 References

1. Netravali, A. N., and B. G. Haskell, *Digital Pictures, Representation and Compression,* Plenum Press, 1988.
2. Musmann, H. G., P. Pirsch, and H. J. Grallert, "Advances in Picture Coding," *Proceedings of IEEE,* April 1985.

3. Beyer, S., "Displacement Predictions for Television Pictures with Minimum Error Variance," *VDI-Verlag,* Duesseldorf, vol. 10, no. 51, 1986.
4. Musmann, H. G., P. Pirsch, and H. J. Grallert, "Advances in Picture Coding," *Proceedings of IEEE,* S. 523 ff, April 1985.
5. Thomas, G. A., "Television Motion Measurement for DATV and Other Applications," *BBC Research Dept. Report No. BBC RD,* 1987/11, 1987.
6. Gilge, M., "Region-Oriented Transformation Coding in Picture Communication," S.136 ff, *VDI-Verlag,* Duesseldorf, vol. 10, no. 128, 1990.
7. Hou, P., and U. Schmitz, "Improved Motion Prediction for Television Pictures," *Rundfunktechnische Mitteilungen* S. 157 ff, Jahrg. 34, 1990/4.
8. Jain, J. R., and A. K. Jain, "Displacement Measurement and its Application in Interframe Image Coding," *IEEE Trans. on Communications,* vol. COM-29, no. 12, S. 1799 ff, December 1981.
9. Girod, B., "The Efficiency of Motion-Compensating Prediction for Hybrid Coding of Video Sequences," *IEEE J. Selected Areas in Communications,* vol. SAC-5, no. 7, S.1140 ff, August 1987.
10. DeVos, L., "VLSI-Architectures for the Hierarchical Block-Matching Algorithm for HDTV Applications," *SPIE,* vol. 1360, S.389 ff, *Visual Communications and Image Processing,* 1990.
11. Funke, J., et al., "System Concept for a Motion Predictor According to the Phase-Correlation Principle with Block Matching," *Dortmunder Fernsehenseminar,* Dortmund, Germany, 1991.

Chapter

5

Video Compression Techniques: Transform Coding

5.1 The Importance of Transform Coding

In the technical press we find countless versions of different coding techniques. Despite the large variety of techniques covered by the media, there is one that crops up regularly, in its different variations, during discussions about future transmission standards: *transform coding* (TC).

There are several reasons for this. Transform coding is a universal method that is well-suited for both large and small bit rates. This fact is not to be taken lightly—in this book we also discuss other methods, such as *vector quantization* and *block-truncation coding,* which are suited to low bit rates (≤ 2 bits/pixel) but cannot provide good picture quality even at high bit rates. By way of comparison, DPCM only gives a good picture quality at higher bit rates (> bits/pixel). Compared with this, TC offers the possibility to code adequately with the same algorithm at a wide range of bit rates. It requires a greater effort to implement TC than simple DPCM, but this disadvantage is more than compensated for by the advantages mentioned.

Furthermore, the subjective impression given by the resulting picture is frequently better than with other methods, because of several possibilities that it offers for exploiting visual inadequacies of the human eye. If the intended bit rate turns out to be insufficient, the effect is seen as a lack of sharpness—which is less disturbing (subjectively) than other coding errors, such as frayed edges or noise with a structure. Only at very low bit rates does TC produce a particularly noticeable artifact: the *blocking effect.*

5.1.1 Transform coding as a coding technique in the frequency domain

In Chap. 1 the very powerful concept of the transform was introduced as a means of representing a signal in the frequency domain. For example, n neighboring pixels $\mathbf{s} = \{s_1,..., s_n\}$ of a picture line in the spatial domain are transformed via a calculation into the frequency domain. The result of this transform is N frequencies $\mathbf{c} = \{c_1,..., c_N\}$:

$$\mathbf{s} \rightarrow \text{transform} \rightarrow \mathbf{c}$$

in which $c_1,..., c_N$ represent the contributions of various spectral components in **s,** whose frequency grows from c_1 to c_N. The scanning spot vector **s** is recovered through the corresponding inverse transform:

$$\mathbf{c} \rightarrow \text{inverse transform} \rightarrow \mathbf{s}$$

Figure 5.1 shows how the transform can be used for data compression: a segment of a picture line is transformed. Subsequently, each coefficient is quantized with a suitable quantizer and finally transmitted as a code word. The quantized coefficients are reconstructed out of the code words and retransformed into the space domain at the decoder.

The payoff from using frequency-based transforms in data compression comes from the fact that natural pictures are composed mostly of low frequencies. Thus, the bulk of the picture information can be

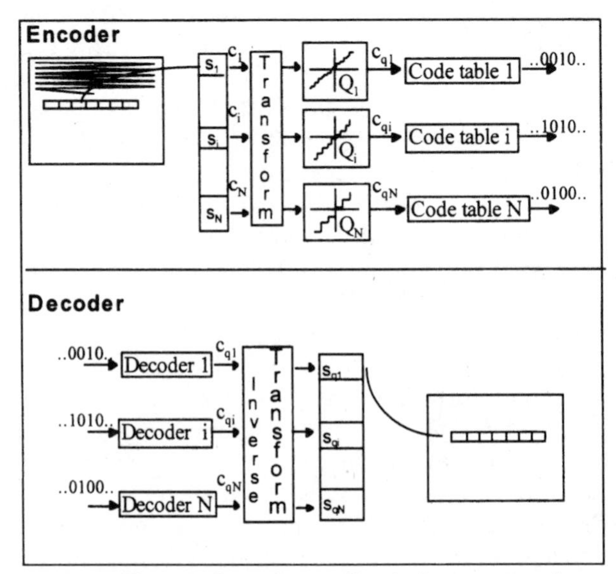

Figure 5.1 The use of transform coding in data compression.

described and coded with a few coefficients that represent the lower frequencies.

5.1.2 A general interpretation of the transform

The interpretation of the coefficients $c_1,\ldots, c_N$ as frequencies is too restricted. In the first instance, the name of the game is to use the transform to concentrate as much information as possible into as few coefficients as possible. There are several transforms that can do it—some better than others—without the coefficients always necessarily representing sinusoidal signal components.

In the example shown in Fig. 5.2, picture segments are transformed out of only two neighboring pixels. The brightness values of different pixel pairs $\{s_1, s_2\}$ are entered as points in a coordinate system with the coordinates S_1 and S_2. As neighboring pixels often have similar brightness, almost all the point pairs find themselves close to the angle bisector. Now we can clearly describe the points entered together with the coordinates of another coordinate system. In Fig. 5.2, a coordinate system has been chosen that has its axis turned in relation to the old axis, and whose axis C_1 now lies precisely on the bisector of the old coordinate system. This coordinate system has been deliberately chosen because most of the pixel pairs lie close to the axis C_1. The point $\{s_1, s_2\}$ can now be described alternatively by a pair of val-

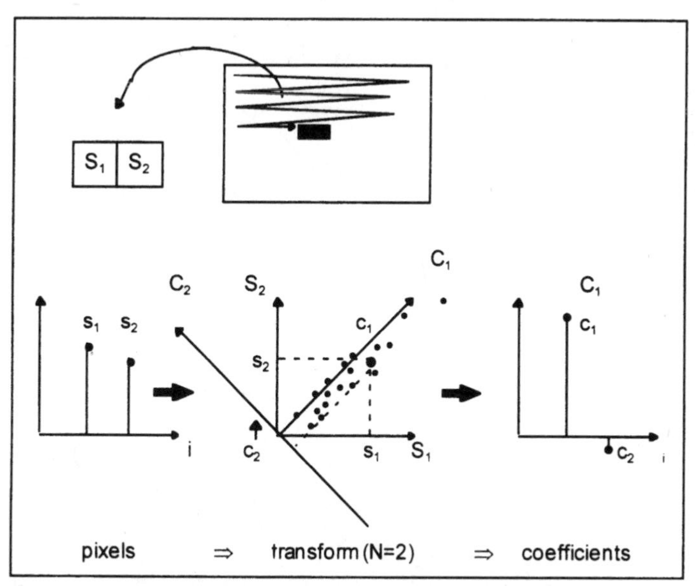

Figure 5.2 Example of a transform that rotates the coordinate axes.

ues $\{c_1, c_2\}$, termed *coefficients*. A transform does exactly that: it describes a data vector of N neighboring points $\{s_1,\ldots, s_N\}$ with skillfully turned coordinates. Representation by other coordinates can be described via a matrix multiplication. The previous transform with $N = 2$, gives

$$c_1 = \frac{1}{\sqrt{2}}\left(s_1 + s_2\right)$$

$$c_2 = \frac{1}{\sqrt{2}}\left(s_1 - s_2\right)$$

or

$$\begin{pmatrix} c_1 \\ c_2 \end{pmatrix} = \frac{1}{\sqrt{2}} \begin{pmatrix} 1 & 1 \\ 1 & -1 \end{pmatrix} \begin{pmatrix} s_1 \\ s_2 \end{pmatrix}$$

$$\boldsymbol{c} = \boldsymbol{A} \times \boldsymbol{s}$$

Of course, we can also recalculate back from the coefficients to the pixels; we turn the coordinate system back around and multiply with the inverse matrix.

$$\begin{pmatrix} s_1 \\ s_2 \end{pmatrix} = \frac{1}{\sqrt{2}} \begin{pmatrix} 1 & 1 \\ 1 & -1 \end{pmatrix} \begin{pmatrix} c_1 \\ c_2 \end{pmatrix}$$

$$\boldsymbol{s} = \boldsymbol{A}^{-1} \times \boldsymbol{c}$$

Because we used an axis which bisected the other axes exactly, it turns out that in this example, $\mathbf{A}$ and $\mathbf{A}^{-1}$ are by coincidence equal.

What has the transform accomplished? If we consider c_1 and c_2, we see that the largest part of the picture information $\{s_1, s_2\}$ is concentrated on one coefficient, namely c_1. It is interesting to see the contribution of c_1 and c_2 in the retransformed pixel pair. In Fig. 5.3, c_1 and c_2 are successively set to zero and the resulting coefficient pairs are retransformed. Obviously c_1 represents the DC part and c_2 the AC part in $\{s_1, s_2\}$:

$$\begin{pmatrix} s_1 \\ s_2 \end{pmatrix} = c_1 \begin{pmatrix} \frac{1}{\sqrt{2}} \\ \frac{1}{\sqrt{2}} \end{pmatrix} + c_2 \begin{pmatrix} \frac{1}{\sqrt{2}} \\ \frac{-1}{\sqrt{2}} \end{pmatrix}$$

$$= \quad c_1 \times \boldsymbol{g} + c_2 \times \boldsymbol{w}$$

The DC part $\mathbf{g}$ and the AC part $\mathbf{w}$ are the *basis vectors* of this simple transform; the values c_1 and c_2 give the weighting of both basis vectors in pixel pairs. Every other transform can be likewise described via its basis vectors: a transform with length N also has N

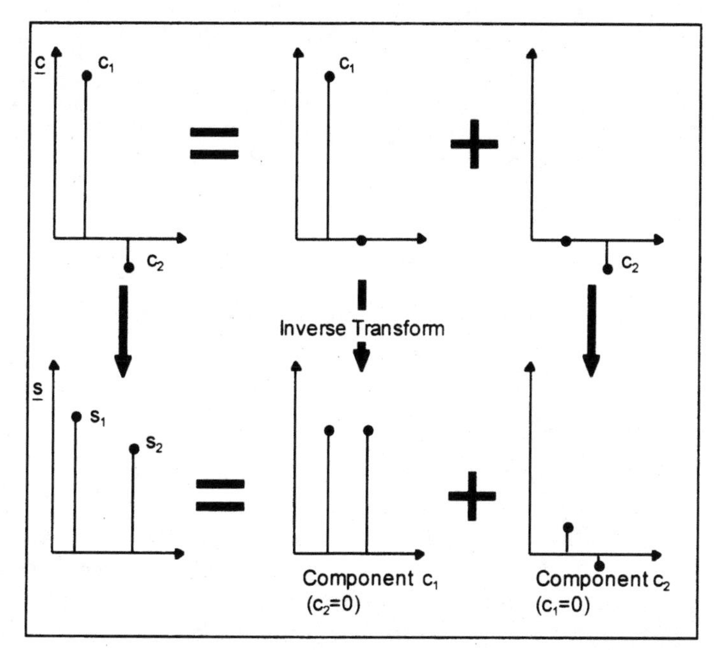

Figure 5.3 Transformation by means of basis vectors; example for $N = 2$.

basis vectors, each with N elements. The N coefficients indicate how strongly represented each basis vector is in the transformed pixel vector $\{s_1,..., s_N\}$. With a good transform, it is possible to construct a pixel vector, adequately and precisely, from a few weighted basis vectors and to represent it with only a few coefficients.

All transforms are described via the same calculation procedure: the multiplication of the pixel vectors $\{s_1,..., s_N\}$ with an $N \times N$ matrix **A.** The various transforms differ only through the entries a_{ij} of their matrices. All meaningful matrices have one common property: they are *orthogonal,* which in essence means:

1. The inverse of a transform matrix is its transposition:

$$\boldsymbol{A}^{-1} = \boldsymbol{A}^T \tag{5.1}$$

2. The sum of the squares of all coefficients equals the sum of the squares of pixels:

$$\sum_{i=1}^{N} c_i^2 = \sum_{i=1}^{N} s_i^2 \tag{5.2}$$

If Eq. (5.2) applies for each transform pair, then it applies most certainly, on average, for a large number of successively performed transforms (*output variation*):

$$\sum_{i=1}^{N} \sigma_{ci}^2 = \sum_{i=1}^{N} \sigma_{si}^2 = N \times \sigma_s^2 \tag{5.3}$$

If the coefficients are quantized, the retransformed signal $\mathbf{s}_q = \{s_{q1}, \ldots, s_{qN}\}$ differs from the original $\mathbf{s}$. The square of the error of the quantized coefficient vector c_q and signal vector s_q is equal because of Eq. (5.2):

$$\sum_{i=1}^{N} \left(c_j - c_{qi}\right)^2 = \sum_{i=1}^{N} \left(s_i - s_{qi}\right)^2 \tag{5.4}$$

Also, this relationship applies on average, i.e., for the average error output of coefficients and signal levels.

5.1.3 Why are coefficients quantized and coded better than pixels?

It might seem that, because of the output variation according to Eq. (5.3), nothing is gained in the transform. The data-reduction capabilities of DPCM lay in the fact that it was not the signal itself which was quantized and coded, but the prediction error. Because this prediction error had a smaller output than the signal, we did not need as many bits.[1] Through the output variation, the sums of the outputs before and after the transform are the same, with the one difference that the outputs of the coefficients are all of different sizes. To make this clear, Fig. 5.4 shows the coefficients c_1, and c_2 observed over several transforms. Whereas the outputs of the signals s_1 and s_2 each represent the signal output ($\delta_{s1}^2 = \delta_{s2}^2 = \delta_s^2$), the output of c_1 is greater than that of c_2:

$$\delta_{c1}^2 > \delta_{c2}^2$$

but

$$\delta_{c1}^2 + \delta_{c2}^2 = {}_{s1}^2 + \delta_{s2}^2 = 2 \times \delta_s^2$$

In order to tell if anything was gained by the transform, an estimate must be made of the data rate R_{PCM}, which is necessary for coding s_1 and s_2. To do this, the approximation already used in previous chapters between the required data rate R, admissible error output s_q^2, and the signal output s_s^2 is applied:

$$R = \frac{1}{6} \times 10 \log \frac{\sigma_s^2}{\sigma_q^2} + k$$

To simplify, $k = 0$ applies in the following for all quantizers. This gives us

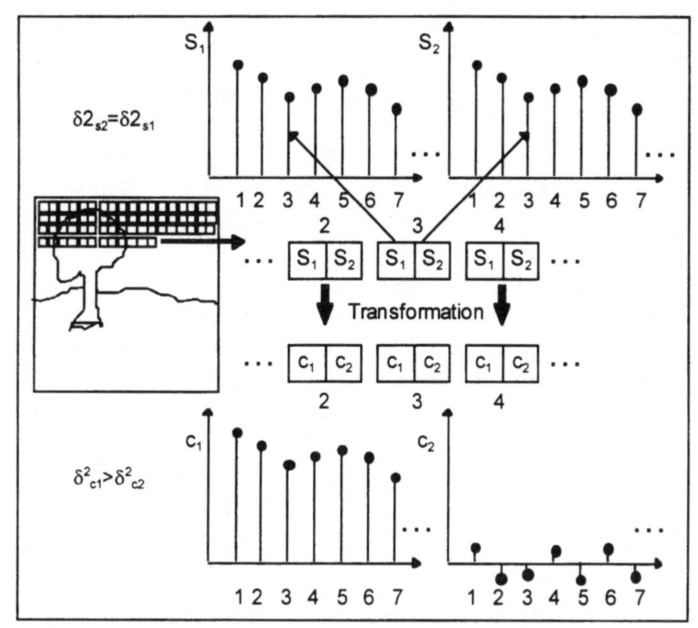

Figure 5.4 Observe the amplitudes of s_1, s_2 and c_1, c_2 in consecutive transformations ($N = 2$).

$$R_{\text{PCM}} = \frac{10}{6} \times \log \frac{\sigma_{s1}^{\ 2}}{\sigma_q^{\ 2}} + \frac{10}{6} \times \log \frac{\sigma_{s2}^{\ 2}}{\sigma_q^{\ 2}}$$

$$= \frac{10}{3} \times \log \frac{\sigma_s^{\ 2}}{\sigma_q^{\ 2}} \left(\frac{\text{bits}}{2 \text{ pixels}}\right)$$

For the coding of the coefficients c_1 and c_2, we need, respectively,

$$R_{\text{TC}} = \frac{10}{6} \times \log \frac{\sigma_{c2}^{\ 2}}{\sigma_q^{\ 2}} \left(\frac{\text{bits}}{2 \text{ pixels}}\right)$$

The same $s_q^{\ 2}$ is also used here. The error in the retransformed pixel vector is, thus, just as large as that for the R_{PCM}-based error, as a result of Eq. (5.4). For $\delta_{c1}^{\ 2} = \delta_{c2}^{\ 2} = \delta_s^{\ 2}$, R_{PCM} and R_{TC} were, in fact, equal. For natural picture contents, $\delta_{c1}^{\ 2}$ is clearly larger than $\delta_{c2}^{\ 2}$ (Fig. 5.4); typical values are $\delta_{c1}^{\ 2} = 1.95\ \delta_s^{\ 2}$ and $\delta_{c2}^{\ 2} = 0.05\ \delta_s^{\ 2}$. Using these two values we get

$$R_{\text{PCM}} - R_{\text{TC}} = 1.68 \text{ bits/(2 pixels)} = 0.84 \text{ bits/(1 pixel)}$$

Obviously, the uneven distribution of the signal output has a favorable effect on the coding. This result can be extended to transforms with transform lengths of $N>2$.

An additional factor must be considered with the calculation of R_{PCM} and R_{TC}: if the error output $\delta_q^{\,2}$ is larger than the coefficient output $\delta_{ci}^{\,2}$, the logarithm will be negative. Of course this does not mean we get any bits for free. In this case the respective coefficient will not be transmitted; hence, the error output will equal the coefficient output. The previous consideration regarding the transform gain is not only valid for the case where all $\delta_{ci}^{\,2}$ are larger than $\delta_q^{\,2}$, but it is also generally the case.[1,2]

The transform is, like DPCM, a method of decorrelation. The concentration on few coefficients is the reward for the decorrelation of the picture contents: although neighboring pixels are similar, there are hardly any similarities between coefficients c_i, c_j of various indices i, j. In fact, the transform that works best is the one with the most concentrated output and, which at the same time, decorrelates the best. Here the degree of output concentration is not only dependent on the transform but also on the character of the picture content. If it receives no correlation, no concentration of output will be reached, even with the best transform.

5.1.4 Which transform is the best?

A problem has recently become clear: not all pictures have the same statistical characteristics. Consequently, the optimum transform is not constant but depends on the momentary picture contents that have to be coded. We could, for example, recalculate the optimum transform matrix for every new frame to be transmitted. Such a transform is named after its inventors, *Karhunen-Loeve transform* (KLT). Although the KLT is optimum, it is not used in practice. It is too troublesome to investigate each new picture to find the best transform matrix. On top of this, the matrix must be indicated to the receiver for each frame, as this must be used in the decoding of the relevant inverse transform. A compromise is made with the *discrete cosine transform* (DCT). This transform matrix is constant and is so suitable for natural pictures that it is sometimes called the quick KLT. Mathematically, this transform looks like this:

$$[\mathbf{A}]_{i,j} = a_{ij} = kj \times \cos\left[\frac{\pi}{N}\left(j - \frac{1}{2}\right)\ (i-1)\right]$$

$$k_i = \begin{cases} \sqrt{\dfrac{1}{N}} & \text{if } i = 1 \\ \sqrt{\dfrac{2}{N}} & \text{if } i > 1 \end{cases} \quad \text{and} \quad i,j = 1,\ldots,N$$

We get the inverse matrix as transponent of **A:**

$$[\mathbf{A}^{-1}]_{i,j} = [\mathbf{A}]_{j,i} = a_{ji}$$

For example, if $N = 4$, we get

$$\mathbf{A} = \frac{1}{2}\begin{bmatrix} 1 & 1 & 1 & 1 \\ a & b & -b & -a \\ 1 & -1 & -1 & 1 \\ b & -\mathrm{a} & a & -b \end{bmatrix} \qquad a = 1.306 \qquad b = 0.541$$

The DCT is a near relative of the *discrete Fourier transform* (DFT), which is widely used in signal analysis. Similar to the quick algorithm for the DFT, FFT offers, like DCT, an algorithm for quick execution of matrix multiplication. In addition to DCT, there are other transforms that are suggested for data compression,[1] such as the Slant transform or the Hadamard transform. The Hadamard transform, for example, uses a slightly weaker output concentration than the DCT. It is easier to realize as all matrix entries are either -1 or 1, and therefore during the transform there are no real multiplications necessary. We can say, however, that of all the transforms, only the DCT has prevailed.

Figure 5.5 shows the basis vectors of a DCT for $N = 16$, and, as an example, the result of a transform. As we can see, the coefficients c_j with increasing index j represent ever increasing frequencies.

5.1.5 Planar transform

The similarities of neighboring pixels are not only line- or column-oriented but also area-oriented. To make use of these neighborhood relations, it is understandable that we would not only like to transform in lines and columns but also in areas. This can be achieved by a *planar transform.* In practice, separable transforms are used almost exclusively. A separable planar transform is nothing more than the repeated application of a simple transform. It is almost always applied to square picture segments of size $N \times N$ and progresses in two steps (Fig. 5.6):

1. All lines of the picture segments are transformed in succession.
2. All rows of the segments calculated in step 1 are transformed.

In textbooks, the planar transform is frequently called a *2D transform.* The transform is, in principle, possible for any segment forms and not just for square ones.[3–5] Figure 5.7 shows results of the planar DCT process.

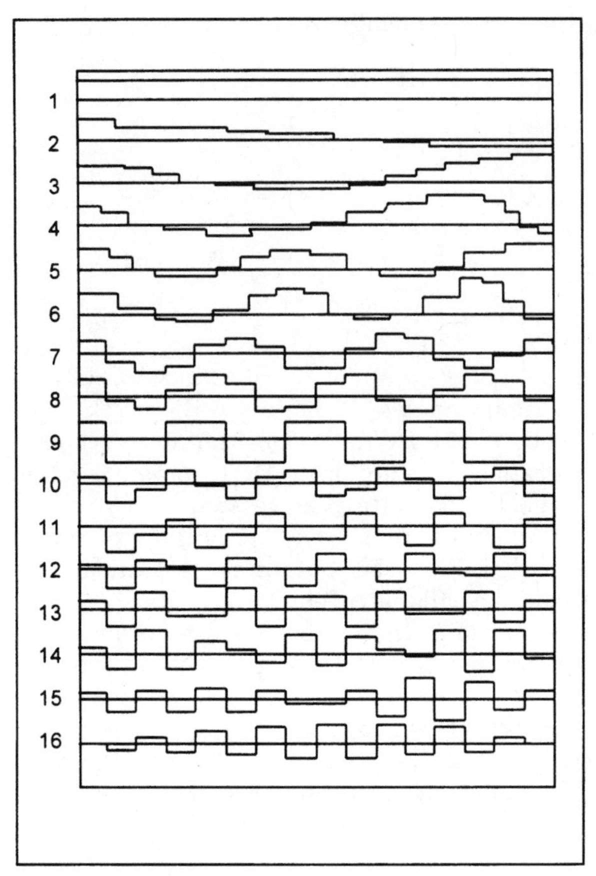

Figure 5.5 The basis vectors of a DCT with $N = 16$.

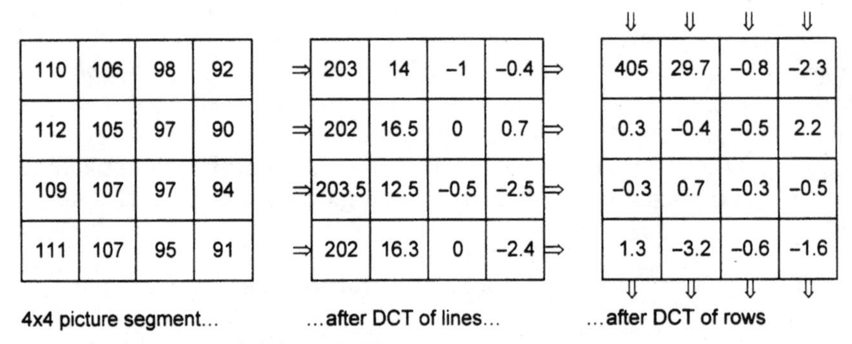

4x4 picture segment...

110	106	98	92
112	105	97	90
109	107	97	94
111	107	95	91

...after DCT of lines...

203	14	−1	−0.4
202	16.5	0	0.7
203.5	12.5	−0.5	−2.5
202	16.3	0	−2.4

...after DCT of rows

405	29.7	−0.8	−2.3
0.3	−0.4	−0.5	2.2
−0.3	0.7	−0.3	−0.5
1.3	−3.2	−0.6	−1.6

Figure 5.6 Processing of a planar DCT.

Figure 5.7 The left half of this picture is transformed with planar DCT (segment size is 32×32).

For a segment of the size $N \times N$, we consequently use $2N$ transforms. The coefficients are now no longer arranged as vectors but as a matrix. The coefficients of the i lines and the j columns are called c_{ij} $(i,j = 1,\ldots, N)$. Each of these coefficients no longer represents a basic vector but a *basic picture.* In this way, each $N \times N$ picture segment is composed of $N \times N$ different basic pictures, in which each coefficient gives the weighting of a particular basic picture. Figure 5.8 shows the basic pictures of the coefficients c_{11} and c_{23} for a planar 4×4 DCT. As we can see, c_{11} represents the DC part. We therefore call it the *DC coefficient*; the others are appropriately called the *AC coefficients.*

The theoretical range of values (not the output) of the coefficients is increased by factor N after the planar transform. If the range of the pixels went from 0,..., 255, the range of c_1 after an $N \times N$ DCT goes from 0,..., $N \times 255$ and all other coefficients from $-(N/2)\times 255,\ldots,$ $(N/2)\times 255$. For practical reasons, the coefficients are frequently scaled down by a factor of $N/2$ after the transform, which is assumed in the following. After being rounded off, the coefficients can then be stored with 9 bits, which actually represents a first, though minor, quantization.

Two aspects must be considered when choosing a suitable segment size. On the one hand it is preferable to use blocks that are as large

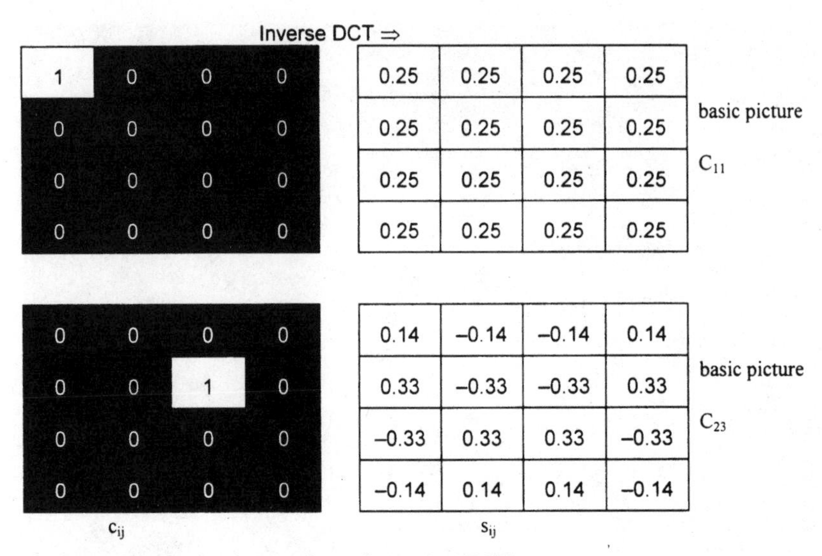

Figure 5.8 The basic pictures for a planar 4×4 DCT.

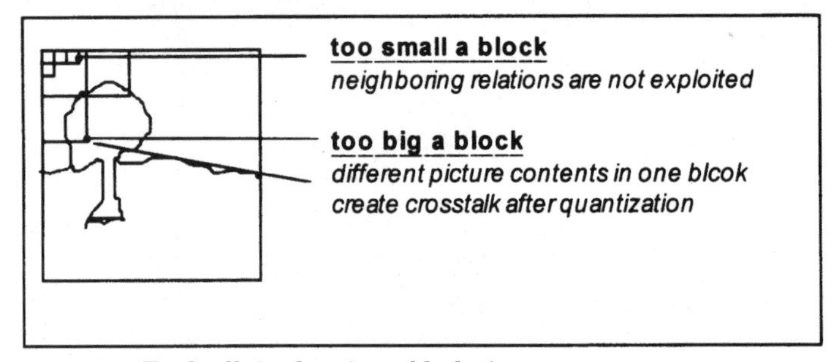

Figure 5.9 Tradeoffs in choosing a block size.

as possible, as this offers a way of building up neighborly relations with pixels—even with those which are remote. But with blocks that are too large there is a strong possibility that in the same picture segment there will be found picture content with both lower activity (for example, a cloudless sky) and higher activity (for example, treetops) simultaneously (Fig. 5.9). But we know that the quantization error of a single coefficient in the space domain is distributed over the entire segment. After quantization and inverse transform, the picture contents reflect the active picture regions in the inactive regions, where the coding errors are clearly visible, e.g., as echoes or blocking (Figs. 5.10 and 5.11). If we use smaller blocks, we can exploit the inadequacies of the human eye more efficiently: segments with higher activity, for example, can be quantized more coarsely than segments with lower activity, because of masking effects. The block size is, therefore,

Figure 5.10 Crosstalk of coding errors within a picture segment.

Figure 5.11 Difference in visibility of errors: low-frequency artifacts on the left, high-frequency artifacts on the right.

a compromise between optimum use of neighborly relations and optimum consideration of local picture characteristics. A further disadvantage of larger segments is that the calculation effort is at least squared with the increased block size. A block size of 8×8 pixels is most commonly used. There are also tendencies for the block size to be modified locally to suit the picture content.[3]

The planar transform of television pictures in the interlaced format is somewhat problematic. In moving regions of the picture, depending on the speed of motion, the similarities of vertically neighboring pixels of a frame are lost, because changes have occurred between the sampling of the two different picture halves. For this reason, the output concentration can be greatly weakened compared to progressively scanned pictures. Well-tuned algorithms, therefore, try to detect stronger movements and to switch to a transform in one picture half (i.e., field) for these picture regions.[4] But the coding in one picture half is less efficient due to the correlation of vertically neighboring pixels being weaker than in the full picture of a static scene. Simply stated, if the picture sequences are interlaced, the picture quality can be influenced by the motion content of the scene to be coded.

5.1.6 Quantizing of coefficients

The most primitive type of quantization is one that simply does not transmit a part of the higher frequency coefficients and codes the remaining coefficients with a fixed number of bits. The complete suppression of coefficients can be interpreted as a quantization with only one quantizing step. This sort of quantization is commonly known as *zonal sampling.*

Although zonal sampling is an effort-saving method of quantizing, there are much more effective methods: for example, different possibilities that take into consideration the targeting of the subjective perception, as well as local picture characteristics. We have previously shown that high frequency parts of signals are not as well perceived as those of low frequency. Further, diagonal resolution has less influence on the subjective picture quality than the horizontal or vertical resolution. This means however that the coefficients to be quantized have a different importance for the impression that the picture gives. Put a different way, quantization errors are less disturbing with coefficients which represent higher and/or diagonal coefficients. To demonstrate this, Fig. 5.11 simulates some quantization errors: three coefficients are given an added disturbance, with a 16×16 DCT. In the left half of the picture, it is the coefficients c_{11}, c_{12}, c_{21}, and in the right-hand half, $c_{10\,10}$, $c_{10\,11}$, $c_{11\,10}$. Although the reconstructed errors in both halves of the picture are, on average, the same size, they are much more noticeable in the left half. This phenomenon justifies the increasingly coarser quantization toward higher frequencies but not the complete suppression of an entire frequency range, as practiced in zonal sampling.

For most applications, linear quantization of the coefficients with subsequent entropy coding has established itself as the norm. As discussed in previous chapters, linear quantizing is easy to realize via a weighting with subsequent rounding off. In combination with entropy coding (Huffman, for example), the linear quantization is almost optimum. If we want to equip each coefficient with an individual quantizer, each coefficient c_{ij} is weighted with a factor w_{ij} and subsequently rounded off. This procedure is shown in Fig. 5.12 with a typical weighting matrix and the coefficient matrix from Fig. 5.6 (scaled with the factor $N/2 = 2$). As we can see, a large number of coefficients are reduced to zero, preferably those which are situated in the lower right corner. The coefficients are subsequently entropy-coded, transmitted, decoded at the receiver, and multiplied with the inverse weighting factor. The coefficient matrix obtained in this way, i.e., faked through quantizing errors, is subsequently retransformed in the space domain to obtain the decoded picture segment. As we can see, the decoded picture segment differs only slightly from the original in Fig. 5.7.

5.2 Implementation Considerations

In the previous sections, the planar transform was introduced as a reversible formula with which square picture segments of $N \times N$ pixels with $N \times N$ coefficients can be calculated. These coefficients describe the picture segment in coordinates that more clearly describe the picture from an information point of view: while the picture information

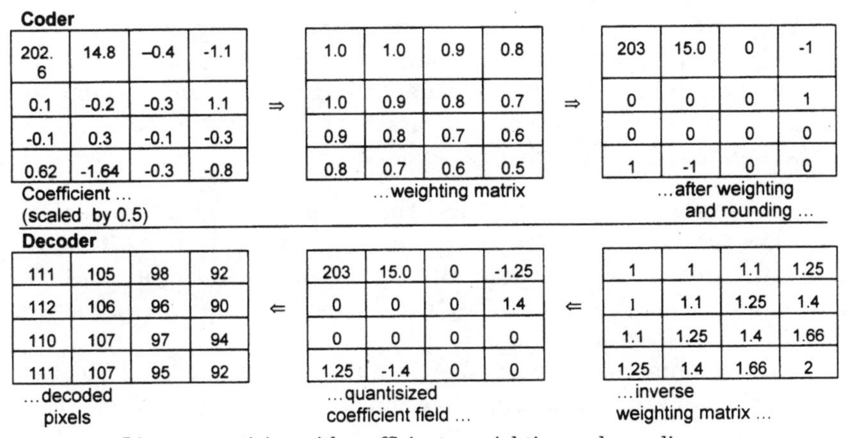

Figure 5.12 Linear quantizing with coefficients: weighting and rounding.

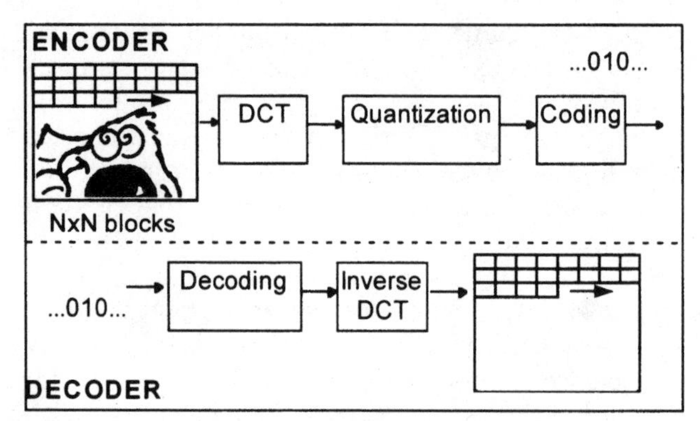

Figure 5.13 Principle of transform coding with DCT.

in the space domain is evenly distributed over all pixels, it is concentrated, after the transform, in just a few coefficients. The degree of this output concentration is a good criterion for selecting the transform.

A schematic diagram of the transform coding is shown in Fig. 5.13. In the simplest form of data reduction, some of the coefficients are not transmitted and the remaining coefficients are quantized with a fixed word length (zonal sampling). This method is very simple, but there are much more efficient quantizing techniques that make use of the following fringe conditions:

- The human eye does not respond well to high frequencies. Thus the coefficients can be quantized with increasing coarseness as the frequencies get higher. If coefficients are not transmitted at all, it is, strictly speaking, a quantization with only one substitute value (namely, zero).
- In regions of the picture with high activity, coding errors are not perceived well, and sometimes they are masked completely. Apart from this, the coefficients (each depending on local picture content) have also different statistical characteristics that can be exploited through modified quantization and coding.

Each of these effects can cancel the other out: in picture regions with low activity the coefficients have a small output, so that we could theoretically save bits. But coding errors are particularly noticeable in precisely these picture regions. A converse argument holds for active picture regions.

There are numerous suggestions for quantizing and coding the coefficients. Most of them fall into two groups: *zonal coding* and *threshold*

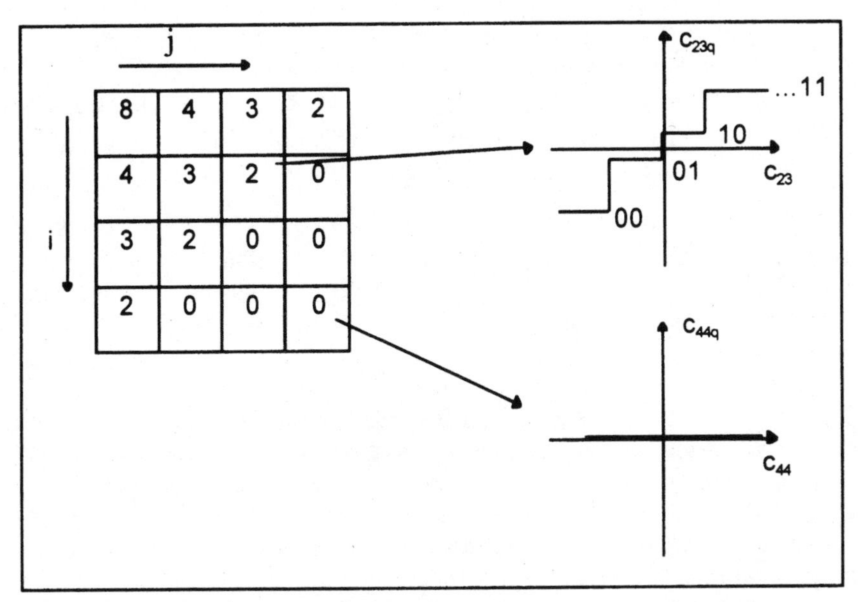

Figure 5.14 Possible bit allocation for zonal coding for a planar 4×4 transform.

coding. With zonal coding,[6–8] coefficients with fixed word length are coded, whereby for each coefficient c_{ij}, a number of bits I_{ij} are made available. The quantization of the coefficients is thereby affected: if only $I_{23} = 2$ bits are estimated for the coefficients c_{23}, the resultant quantizer has $2^2 = 4$ quantization steps (Fig. 5.14). The DC coefficient is mostly quantized with 8 bits, and the AC coefficient, depending on the frequency, with 0,..., 8 bits. With adaptive zonal coding, each picture segment is quantized according to its activity level, and these various activity levels are then assigned to one of M classes of A_k (k = 1,..., M) according to an activity criterion. For each class there is a different bit allocation I_{ij} (k). In Eq. (5.2) for example, the complete picture is transformed, segment by segment, and each segment put in one of four activity classes A_k (k = 1,..., 4). The activity criterion applied is the sum of the squares of all coefficients within a segment; the bit allocation is made according to purely statistical grounds, without consideration of masking effects. A simplified classification into eight classes can be found in Ref. 8, whereby the local construction of images is taken into consideration.

An advantage of zonal coding is that during coding with fixed word lengths, the bit rate can be simply controlled and held constant for each frame. Zonal coding also simplifies error protection. But this method still has some problems, including

- The coding with fixed word length is not optimum with regard to the data rate,[9] especially when linear quantizers are used.
- A code word must be transmitted for each coefficient, even when it is quantized to zero. Normally, most of the coefficients are quantized to zero; therefore such coding no longer makes sense.

In threshold coding, the coefficients are coded with a variable word length after linear quantization. For this reason, far more quantizing steps can be allowed for linear quantizing, whereas with entropy encoding the bit rate depends far less on the number of steps and more on the resolution (step size). Thus, a much larger dynamic range can be chosen for the linear quantizer than with zonal coding. Furthermore, a code word does not have to be transferred for each zero, and after quantizing, neighboring zeros are grouped and coded with a common code word. In some methods, the lowest quantizing step of the quantizer is raised slightly. More of the small coefficients that lie under the threshold are thus quantized to zero. Psychovisually, this additional trick is not justifiable, as the visibility of quantizing errors depends more on the spatial frequency of the coefficients c_{ij} than on their amplitude.

Threshold coding (a much quoted version can be found in Ref. 10) has above all established itself in practice. This method and its different variations are the basis for many standard suggestions for interframe coding[11] and hybrid coding.[12–14] They cover linear quantizing, entropy coding of coefficients, and buffer control. Furthermore, the masking effects mentioned previously can be taken into account. In order to make use of the similarities of neighboring pixels during data compression, we can apply the DCT to the difference picture of a motion-compensated DCT. This method, known as *hybrid DCT,* is introduced in Chap. 6.

5.2.1 Intraframe transform coding

In the previous section, it was suggested that each coefficient c_{ij} is equipped with a linear quantizer with an individual step height s_{ij}. The coefficients are thereby multiplied by a weighting factor and rounded off. The size of the weighting factor w_{ij} and, thus, the step height s_{ij} of the linear quantizer ($s_{ij} = 1/w_{ij}$), depends on how strongly visible a quantizing error of the coefficients c_{ij} will be. As already demonstrated in Fig. 5.11, quantizing errors with frequently occurring coefficients are not so disturbing. They can therefore be more coarsely quantized. Figure 5.15 shows a typical weighting matrix for an 8×8 transform.

→ j

↓ i

1.00	0.73	0.80	0.50	0.33	0.20	0.16	0.13
0.67	0.67	0.57	0.42	0.31	0.14	0.13	0.15
0.57	0.62	0.50	0.33	0.20	0.14	0.12	0.14
0.57	0.47	0.36	0.28	0.16	0.09	0.10	0.13
0.44	0.36	0.22	0.14	0.12	0.07	0.08	0.10
0.33	0.23	0.15	0.13	0.10	0.08	0.07	0.09
0.16	0.13	0.10	0.09	0.08	0.07	0.07	0.08
0.11	0.09	0.08	0.08	0.07	0.08	0.08	0.08

Figure 5.15 Typical weighting matrix for an 8×8 transform; it is difficult to distinguish between the original and decoded picture.

While the DC coefficient c_{11} represents the DC part within a picture segment, the remaining so-called AC coefficients are responsible for the contrast and the activity within a segment. As coding errors are not perceived so well in segments with higher picture activity, a smaller weighting factor can be used in such segments. After the transform, each segment is typically allocated to one of M activity classes. A weighting factor a_k ($k = 1,\ldots, M$), which is as small as the measured activity is large, belongs to each activity class A_k. In addition to w_{ij}, each AC coefficient is also weighted with a_k:

$$c_{ij}' = \begin{cases} c_{ij} \times w_{ij} & \text{for } i = 1, j = 1 \\ c_{ij} \times w_{ijak} \times a_k & \text{for all others} \end{cases}$$

A problem occurs with the definition of suitable activity classes. (Some points concerning this can be found in Refs. 17 and 18, although a classification according to statistical and not a psychovisual point of view is suggested there). Taken literally, it must be determined whether structured or unstructured picture activity exists and which local direction the existing structure has, for in this case various different perception characteristics of the eye come into play. But

202.6	14.8	-0.38	-1.13
0.12	-0.24	-0.26	1.11
-0.13	0.30	-0.13	-0.26
0.62	-1.64	-0.30	-0.77

		a_k
A_1	0...10	1.0
A_2	10...20	0.66
A_3	20...50	0.5
A_4	50...	0.4

Figure 5.16 Allocation into groups by largest AC coefficient value; example with a 4×4 segment and four groups.

even quite simple classifiers give adequate results. Some algorithms search in this way for the largest AC coefficients in the transformed segment and use their amplitude to select the appropriate class.[19] Figure 5.16 shows an example of a division into four classes. The momentary picture segment is allocated to class 2, as the amplitude of the largest AC coefficient is between 10 and 20. These classes must be communicated to the receiver, as the inverse weighting must be done with $1/a_k$. For four classes, an additional 2 bits must be transmitted per segment. The common weighting of all coefficients with a single factor a_k is a simple way of exploiting masking effects. More effective still is to develop a weighting matrix for every activity class, using subjective tests.

We can easily imagine that the number of bits to be transmitted per segment depends on the local characteristics of the picture content. In picture regions with poor contrast, few bits are frequently used because the DC part contains almost the complete picture information; in active picture regions, despite certain savings resulting from masking effects, more bits must be transmitted. As almost all channels transmit a constant data rate, these fluctuations must be balanced by a data buffer. This buffer supplies the channel with a constant output rate, even when the data rate fluctuates at the buffer input. Each buffer has a finite size (e.g., 1.6 Mbit in Ref. 12). Thus the danger of overflow arises with each buffer, in the case of complicated pictures. To avoid this, a buffer control is introduced. As the bit rate can be easily controlled by the step height of the quantizer, during entropy coding, a third weighting f is carried out, which depends on the momentary buffer contents level B: the fuller the buffer is at any given moment, the lower is the weighting factor. The weighting for the coefficients c_{ij} are finally

$$c_{ij}' = \begin{cases} c_{ij} \times w_{ij} & \text{for } i = 1, j = 1 \\ c_{ij} \times w_{ij} \times a_k \times f(B) & \text{for others} \end{cases}$$

Because the decoder does not know the contents level B of the coder, the contents level must be communicated to the decoder. As the buffer level changes only slowly, it is adequate (for many applications) to read and transmit the level at intervals of several picture segments. The factor f is held constant within these intervals. Figure 5.17 shows a possible course of the weighting factor f as a function of the buffer contents. The following points stand out with respect to this curve:

1. If the buffer is exactly half full, there is no weighting ($f = 1$).
2. If the buffer is less than half full, a finer quantization takes place. For $f > 1$, the quantizing steps are finer rather than coarser: this produces more data.
3. The curve is very flat over a large region, so that the buffer is best exploited. This way it is possible for bits that are in regions of lower picture activity to be saved and to be used for more critical regions.

After weighting with w_{ij}, a_k, and f, all coefficients are rounded off; the sign is kept and the value is rounded to the nearest whole number. Following this rounding, well over half of the coefficients become zero (Fig. 5.18). In contrast to zonal sampling, the position of these

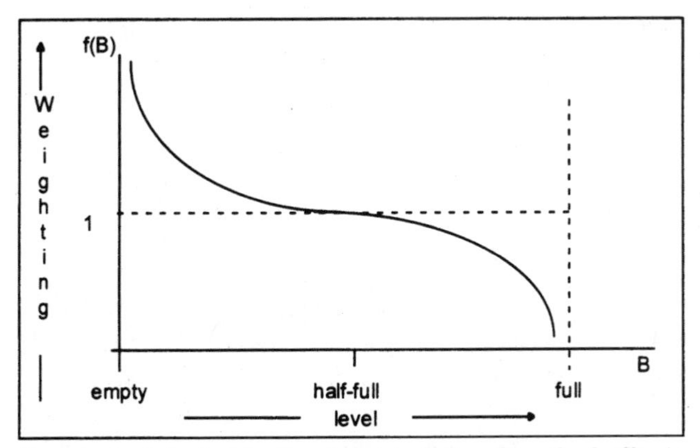

Figure 5.17 A typical buffer characteristic: when the buffer is full, the quantizer step size has to be increased.

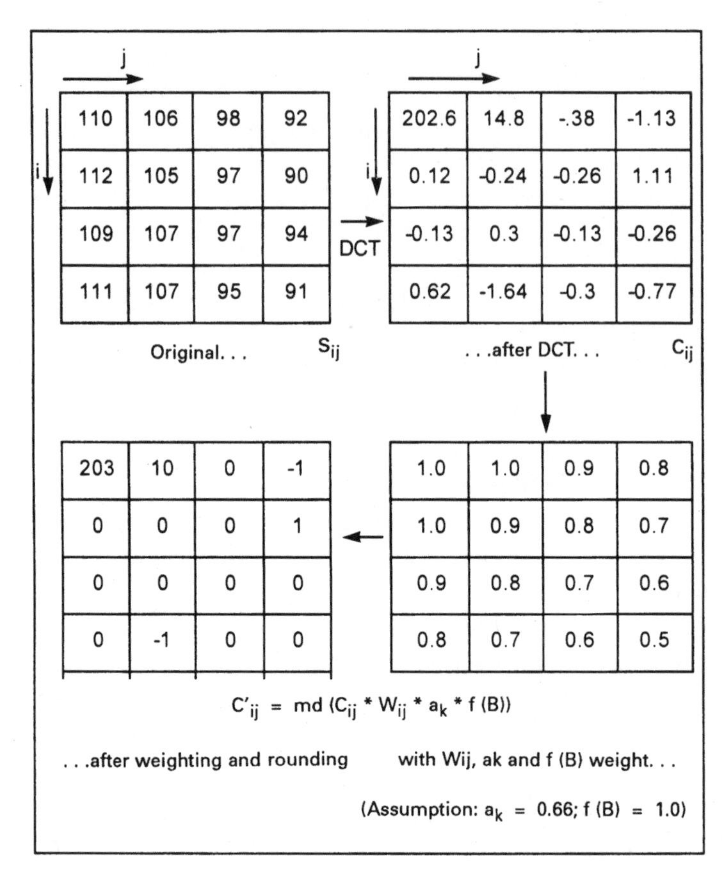

Figure 5.18 Transformation and quantization example with a 4×4 segment.

zeros in the coefficient matrix is unknown to the receiver, which is why they must be coded. To avoid this, each zero has to be given a code word which is too short, and this coding variant, called *run-length coding,* has become well established.

5.2.2 Coding of the coefficients

Next, the coefficients c_{ij} are read out of the coefficient matrix in a zigzag fashion and then put into a vector $\mathbf{v} = v_1,\ldots, v_N{}^2$ of length $N \times N$ (Fig. 5.19). This so-called *zigzag scanning* of the coefficients leads approximately to a sorting of the coefficients in vector **v** according to their amplitude: the elements v_i become smaller on average with increasing index. Values of $v_i > 10$ are mostly only found in the first three to five components of **v.** With subsequent components, only smaller amplitudes, frequently separated from each other by a long

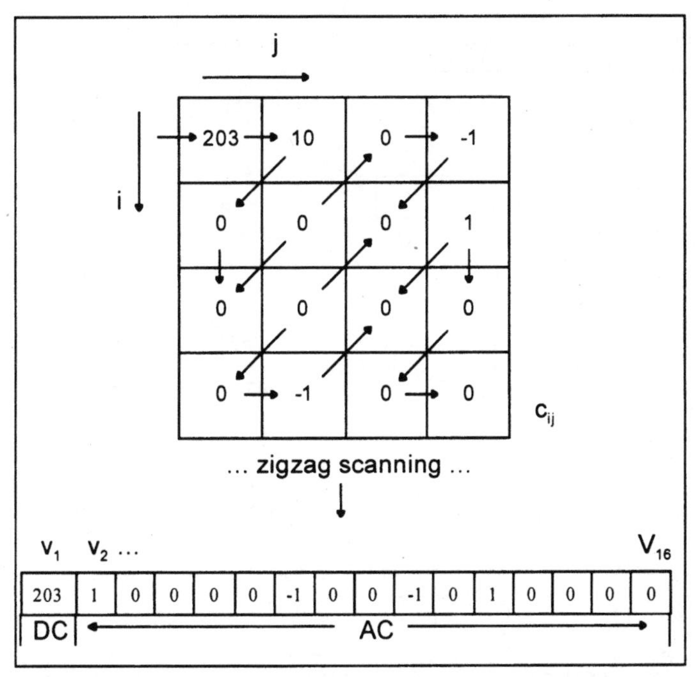

Figure 5.19 Zigzag scanning; the coefficients are sorted into a vector.

succession of zeros, are found. A long succession of zeros is usually found at the end of the vector.

For coding, vector **v** is processed from v_1 to $v_N{}^2$. First, the DC coefficient (0,..., 511) is coded with 9 bits. Next, the zero sequences are combined with the subsequent coefficients. Figure 5.20 shows the sequence 0, 0, 0, 0, -1 resulting in the event (number of zeros = 4; amplitude = -1). A suitable code word is then chosen for this event from a code table.

The most frequent combinations are composed of short successions of zeros and small amplitudes. Therefore, with an 8×8 DCT, it is sufficient to restrict the Huffman code to lengths and amplitudes from 0 to 15. If, however, an amplitude or succession of zeros is not represented in the table, the length and amplitude are coded with a fixed word length. In this case, both code words must be given a clear prefix so that the decoder knows that the code table may not be used for the decoding of this pair of values. A long succession of zeros is mostly found at the end of vector **v.** In this case the decoder is sent a clear end-of-block code word. It then knows that only zeros follow, and the transfer of these segments has come to an end. The code table could be designed with the Huffman algorithm,[12] for example, if the aver-

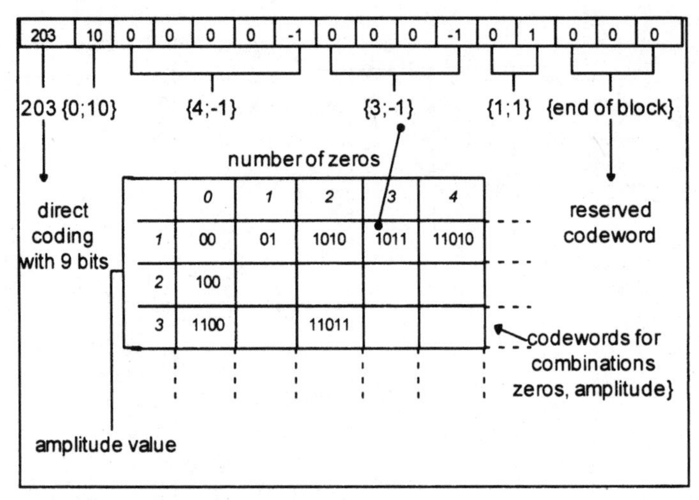

Figure 5.20 Coefficient coding with a Huffman table.

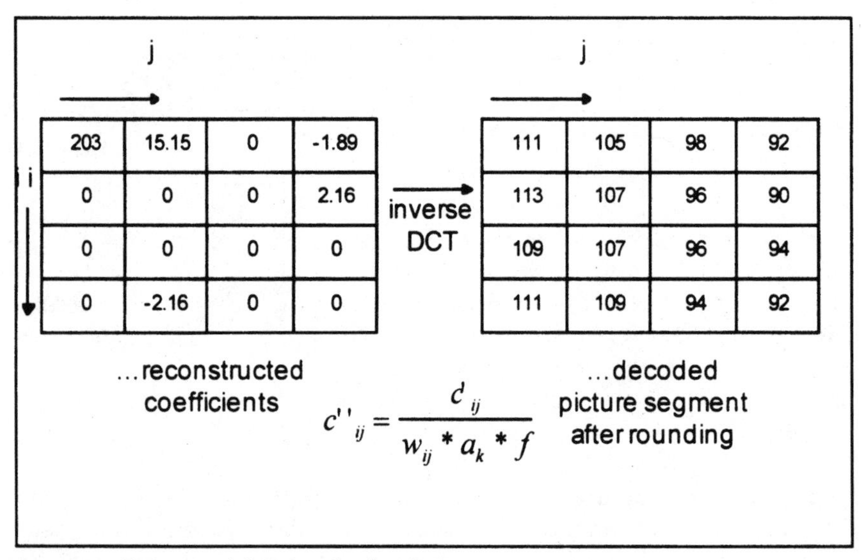

Figure 5.21 Inverse weighting at the decoder.

age occurrence of all possible combinations of zeros and amplitudes is known, as well as the prefix and end-of-block code words.

This coding is shown in Fig. 5.20 for a hypothetical 4×4 coefficient field. As additional information, the sign must be transmitted for each coefficient c_{ij}>0, and for each segment, the class a_k and, if necessary, the buffer contents level B must also be transmitted. Figure 5.21 shows, finally, the reconstructed coefficient field and the decoded pic-

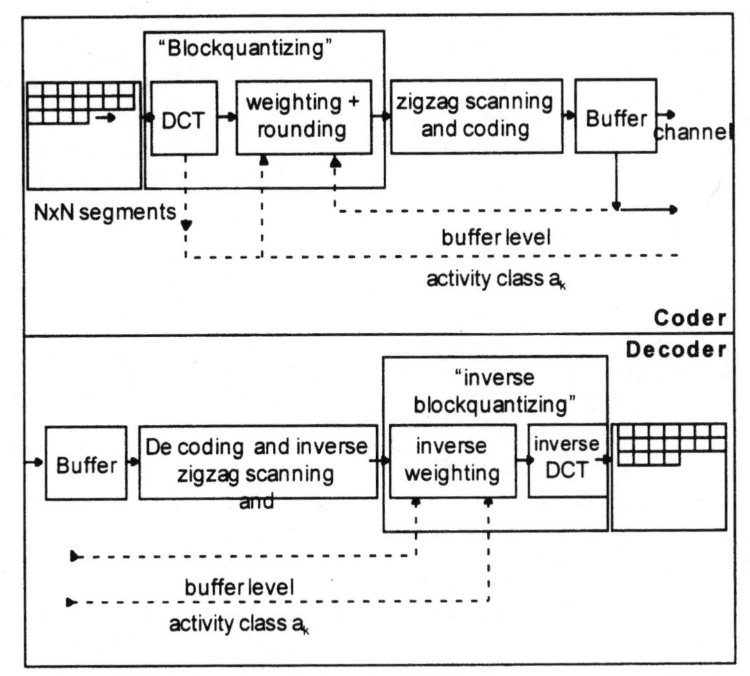

Figure 5.22 Block diagram of a complete DCT codec.

ture segment. Figure 5.22 shows a block diagram of the total codec, as described previously.

5.2.3 Additional considerations

The question arises: How can the foregoing coding information be justified in information theory. The intention and purpose of the transform was actually to obtain uncorrelated coefficients, so that a common coding of several coefficients, called *extension,* could be ignored. But basically, the combination of zero successions and amplitudes described previously is precisely such an extension. Such a case is addressed in Chap. 3, for which an extension of memoryless sources is advantageous in those cases where the entropy of the source event is less than 1 bit. That applies exactly to the most frequently occurring coefficients after quantizing.

We could assume that the differences of the amplitude distributions of different coefficients are not taken sufficiently into consideration: the higher the index i is, the smaller, on average, are the amplitudes and the longer, on average, are the successions of zeros. Despite this, the source events are always coded (length, amplitude) with the same table, regardless of whether this event is at the beginning or at the end of the vector **v**. It is true that the efficiency of this coding could be increased, if two coding tables were used, in which the one that is in the first half of **v** is used, and the other in the second. Different coding tables could be used in the same way for the different buffer contents levels and activity classes. Each case should, however, be investigated to see if the additional effort is justified. In general, the algorithms work with only one coding table.

With color signal coding, the color difference signals u and v, and i and q, must be coded as well as the luminance signal. As opposed to brightness, the color resolution plays a smaller role for the subjective impression of the picture. The color information is therefore often low-pass–filtered and subsampled before the actual coding. The input data rate of each color component is typically four to eight times smaller than the luminance. The coding of color and luminance is carried out using the same algorithm; the color information is, however, often quantized more coarsely.

5.2.4 Interframe transform coding

With the algorithm described previously, compression factors of approximately 8 can be achieved while maintaining good picture quality. To achieve higher factors, we have to exploit similarities between successive frames. The nearest approach to this is the extension of the DCT in the time dimension. These *cubic* transforms are, for example, investigated in Ref. 21, and are suggested for use in a digital VCR.[22] The disadvantages of this approach are the increase in calculation effort and above all the increasingly higher memory requirement: for an $8 \times 8 \times 8$ DCT, at least seven frame memories would be required. Much simpler is the *hybrid DCT*, which also codes pictures with moving objects more efficiently. This method comprises, almost exclusively, a motion-compensated DPCM, as described in Chap. 4; instead of transferring each picture individually, the motion-compensated difference of two successive frames is coded. The unique aspect of this hybrid DCT is that the difference frame is transform-coded. Figure 5.23 gives a block diagram of the hybrid DCT; the block quantization contains the DCT with connected weighting and rounding off (see Fig. 5.22). Because the coder in the feedback loop needs the quantized difference frame in the untransformed form, an inverse DCT must be carried out before the summation point. Thus, from the point of view of the DPCM coder, a quite normal quantization has taken place.

The coder also needs the difference frame in the untransformed state. For this reason an inverse DCT is also carried out at the DPCM

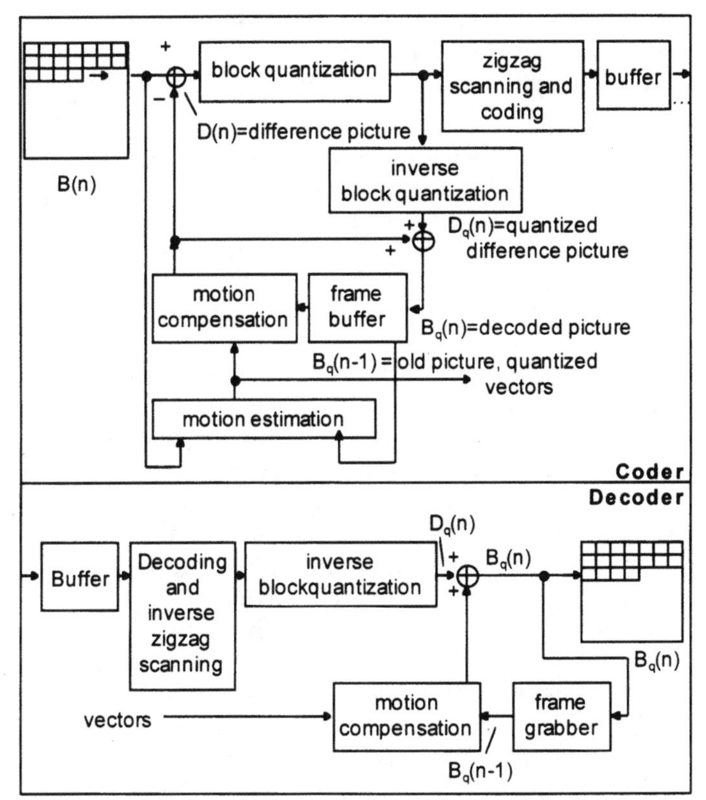

Figure 5.23 Hybrid DCT: use of transform coding on difference pictures of a motion-compensated DPCM.

decoder so that it does not notice that the quantization has taken place in the frequency domain.

Although the hybrid DCT represents the most popular method at present, there are recurring doubts about the efficiency of a DCT if applied to difference frames.[23] Basically, the frames have already been decorrelated via the temporary DPCM, so why is another transform necessary? It is a fact that we observe a much weaker output concentration during a difference frame transform. There are many other suggestions for quantizing difference frames which work without a transform. The DCT could nevertheless once again predominate. Obviously the decorrelation characteristics are not the only advantages of transform coding:

- As shown, the characteristics of the visual system can be used to advantage with the DCT. The coding artifacts from transform coding (unsharpness, for example) are often less disturbing than artifacts produced by other techniques. The bit rate can also be controlled better with the quantizing and buffer concept described previously.

- In the DPCM system, all regions of the picture are refreshed at regular intervals. The DPCM loop is periodically interrupted and the region that is to be refreshed is transform-coded directly and not as a difference. The DCT is well suited for intracoding.

5.2.5 Intraframe DCT or hybrid DCT?

There is no doubt that higher compression factors can be achieved with hybrid DCT than with intraframe DCT. The compression rate, however, is not the only selection criterion. For some applications the possibility must exist for successive frames to be decoded independent of each other. If we want to cut scenes together, for example, which have been coded with a hybrid method, the sequences must be decoded, then cut, and coded again. This represents a lot of effort and also causes additional picture quality penalties resulting from the multiple coding. A further problem is the quick search speed of digital recorders. The picture sequence can only be read in fragments that must be decoded independently. It would be almost unthinkable for a (quicker) rewind of hybrid coded sequences, as the previous—instead of the next—frame is necessary for decoding the frame information.

For all these reasons, intraframe coding is preferred to hybrid coding, wherever picture sequences are processed retrospectively. A typical application is digital recording in studios, but quick-search functions should also work properly in consumer equipment. The application areas for hybrid coding lay mainly where picture sequences are transmitted for the end user, as, for example, within digital television or videophones.

To combine the advantages of hybrid DCT and interframe DCT, we can code whole frames with interframe coding at fixed intervals using hybrid techniques. Single pictures cannot be decoded independently in this way, but nevertheless groups of pictures can. This idea is used in the MPEG standard, which is presented later in this book.

5.2.6 Typical coding errors caused by quantizing and transmission

As with every other coding method, transform coding is affected by quantizing and transmission errors. Every coefficient represents the weighting of a basic frame. When this coefficient is quantized, the related basic frame (weighted with the quantizing error) is laid on top of the original segment. A quantizing error of the DC coefficients shows itself as the superimposition of a DC value in the space domain. The quantizing error of a single coefficient spreads itself in the space domain over the entire segment. The following disturbances show up in the picture:

- Discontinuities occur in the segment borders because of the independent coding of neighboring segments. This can cause block structures (blocking) in regions of the picture with low activity.
- Even weak block effects can suddenly become visible if the scene moves, relative to the block-raster (dirty-window effect).
- If the high-frequency coefficients are too weakly weighted, the picture can visibly lose sharpness.
- Echoes (ringing) can occur in areas of high contrast.
- With the hybrid DCT, coding errors are diffused in the time dimension. This frequently causes unsharp movement. Coding errors in neighboring picture segments are thereby exported, due to motion compensation.

Figure 5.24 shows the decoded picture of an intraframe transform coding for two different compression factors.

The influence of channel errors is restricted to the transmitted signal containing the error(s), provided that the start of the next segment can be detected. That may be easier to realize for zonal coding than for threshold coding, as the number of bits to be transmitted per segment is fixed with the former method. With channel errors (unlike with quantizing), we have no influence over which coefficients are more strongly faked than others. In practice, we must therefore make sure that the coefficients receive appropriate channel protection. We can, thus, be reasonably sure that—judged visually—all coefficients make a subjective contribution to the decoded picture, which is no longer dependent on the spatial frequency. The most important coefficients are then those with the largest amplitudes and must therefore be better protected than the remaining coefficients. In the first approximation, these are the low-frequency coefficients. There are also suggestions to sort the coefficients according to size, before the entropy coding, in order to simplify the subsequent channel coding.

5.2.7 Practical considerations

Through the availability of ever faster and cheaper hardware components, it has recently become possible to put a complete transform coding system on a few microchips and offer them at an affordable

(*a*)

(*b*)

Figure 5.24 Decoded picture of intraframe DCT with compression factors of (*a*) 8 and (*b*) 20.

price. But to open the way toward wide introduction of picture coding in the field of picture communication, more must be done than just the development of algorithms. It must be ensured that coders and decoders understand each other. It is, therefore, a major challenge to define transmission standards which

- Represent the technical state of the art
- Leave room for future developments
- Are suitable for codecs of different complexity and for different bit rates
- Are suitable for different requirements

The DCT techniques have established themselves among today's standards and are also the candidates with the best prospects for future standards. Although the basic concepts of DCT coding are well established, there are still some special features and refinements that are worth reporting. In the next chapter we will introduce some of the most important standards and compare their similarities and differences.

5.3 References

1. Netravali, A. N., and B. G. Haskell, *Digital Pictures, Representation and Compression,* Plenum Press, 1988.
2. Jayant, N. S., and P. Noll, *Digital Coding of Waveforms,* Prentice Hall, New York, 1984.
3. Chen, C-T, "Adaptive Transform Coding via Quadtree-Based Variable Block Size DCT," *ICASSP,* Glasgow, 1989, pp. 1854–1856.
4. deWith, P. H. N., "Motion-Adaptive Intraframe Transform Coding of Video Signals," *Philips J. Res.,* vol. 44, 1989, pp. 345-364.
5. Gilge, M., "Region-Oriented Transform Coding in Picture Communication," *VDI-Verlag, Advancement Report, Series 10,* 1990.
6. Pratt, W. K., W-H Chen, and R. Welch, "Slant Transform Image Coding," *IEEE Trans. Comm.,* vol. 22, August 1974, pp. 1075–1093.
7. Chen, W-H., and C. H. Smith, "Adaptive Coding of Monochrome and Color Images," *IEEE Trans Comm.,* vol. 25, pp. 1285–1292.
8. Zhang, W-J., S. Y. Yu, and H. B. Chen, "A New Adaptive Classified Transform Coding Method," *IEEE, ICASSP* Glasgow, 1989, pp. 1835–1837.
9. Chen, W-H., and W. K. Pratt, "Scene Adaptive Coder," *IEEE Trans. Comm.,* vol. 32, March 1984, pp. 225–232.
10. JPEG, *Digital Compression and Coding of Continuous-Tone Still Images,* Draft ISO 10918, 1991.
11. "Digital Coding of Component Television Signals for Contribution—Quality Applications in the Range 34–45 Mbit/s," *ETS/NA Coding Expert Group,* February 1991.
12. "Video Codec for Audio Visual Services at $P \times 64$ Kbits/s," *CCITT Recommendation H.* 1990, p. 261.
13. WG11, "Coding of Moving Pictures and Associated Audio for Digital Storage Media at up to about 1.5 Mbits/s," *ISO-IEC/JTC1/SC29/MPEG91,* November 1991.

14. Gimlett, J. I., "Use of Activity Classes in Adaptive Transform Image Coding," *IEEE Trans. Comm.,* vol. 23, no. 7, July 1975, pp. 785–786.
15. Strickland, R. N., "Experiments on the Use of Local Statistics for Adaptive Image Processing," *Proc. IEEE, ASSP,* vol. 3, March 1984.
16. Haghiri, M., and P. Denoyelle, "A Low Bit Rate Coding Algorithm for Full Motion Video Signal," *Signal Proc. Image Communication,* no. 2, 1990, pp. 187–199.
17. Natarajan, T. R., and N. Ahmed, "On Interframe Transform Coding," *IEEE Trans. Comm.,* vol. 25, November 1977, pp. 1323–1329.
18. Onishi, K., et al., "An Experimental Home-Use Digital VCR with 3D-DCT and Superimposed Error Correction Coding," *IEEE, ICCE,* Chicago, June 1991, pp. 182–183.
19. Strohbach, P., "Tree-Structured Scene Adaptive Coder," *IEEE Trans. Comm.,* vol. 38, April 1990, pp. 477–486.
20. Gilge, M., "Region Oriented Transform Coding in Picture Communication," *VDI Advancement Report,* series 10, no. 128.
21. Bage, M. J., "Interframe Predictive Coding of Images Using Hybrid Vector Quantization," *IEEE Trans Comm.,* vol. 34, April 1988, pp. 411–415.
22. Kauff, P., P. Stammnitz, and R. Schaefer, "A Codec for the Data Reduced Magnetic Recording of HDTV Signals in Studios," *Dortmund Television Seminar,* LS-NT, Uni. Dortmund, October 1991, pp. 184–189.
23. de With, P. H. N., and S. M. C. Borgers, "On Adaptive DCT Coding Techniques for Digital Video Recording," *IERE Proc. 7th Inter. Conf. Video, Audio and Data Recording,* York, IERE, London, March 1988, pp. 199–204.

Chapter

6

Video Compression Techniques: Vector Quantization

6.1 Introduction

In the previous chapters we have looked at two coding techniques that, to date, are the best researched and therefore the most understood: DPCM and transform coding. Besides these two techniques, there are other methods which could become serious competition for the current favorites. One of these methods is *vector quantization* (VQ), which has been researched intensively since 1980. If the importance of a coding technique was measured by the number of years it has been in existence, VQ transform coding would be at the forefront.

There are many reasons for this interest, but these reasons are not always to be found in the compression results achieved to date. The thinking and the mathematical principles behind VQ are closely related to those of pattern recognition, cluster analysis, and channel coding, so that interest comes not only from the side of source coding. There are geometric interpretations for the various types of VQ, which give this subject a certain attraction, not in the least of which is that it can be shown that VQ is theoretically superior to DCPM and transform coding if we allow unlimited effort in the realization of the methods mentioned. It is a fact that transform coding is only a special case of VQ, even if it would be nonsensical to implement it as VQ. On a practical level, the most promising use of VQ is in transform coding as a hybrid method in combination with DPCM or subband coding.

6.1.1 Review: scalar quantizers

The quantizers, as they have been applied in the previous chapters, i.e., to single values and not vectors, are called *scalar quantizers.* Scalar quantizers are special cases of vector quantizers, as, indeed, a scalar is also a special case of a vector. A scalar quantizer is completely described by the naming of its substitute values. Preliminary to the quantization is the finding of a substitute value s_q for a given input value s, which lies next to the input value. Each substitute value is coded with a distinct code word. The code word length can be the same for all substitute values (fixed-length coded quantizer), or the length may vary from substitute value to substitute value (entropy-coded quantizer).

In special cases, we try not to exclusively minimize the distance between the input value and the substitute value but to choose, simultaneously, a substitute value with a code-word length that is as short as possible (*entropy-constrained quantizer*). In this case the decision threshold must also be given to the substitute value. With these quantizers, an input value is not necessarily quantized with the nearest substitute value but possibly with a somewhat distant substitute value, if it is coded with a shorter code-word length. The advantage that can be gained from this is, unfortunately, often small.

Furthermore, we differentiate between linear quantizers and nonlinear quantizers, which distinguish themselves through equidistant and nonequidistant substitute values. An important difference between linear quantizers and nonlinear quantizers is that the substitute value which fits in the first instance through weighting and rounding off can be simply calculated, whereas in the latter case this substitute value has to be searched for in a table (with the exception of compander techniques).

6.2 Vector Quantizers

With a vector quantizer, we quantize several values at the same time with an input vector of

$$\mathbf{s} = (s_1, \ldots, s_N)$$

with N values. These values can be, for example, pixels lying next to each other. Figure 6.1 illustrates a case in which $N = 4$ neighboring pixels. Whereas we described the scalar quantizer with its substitute values, with the vector quantizer we have *substitute vectors*

$$\mathbf{c}_i = (c_i 1, \ldots, c_i N)$$

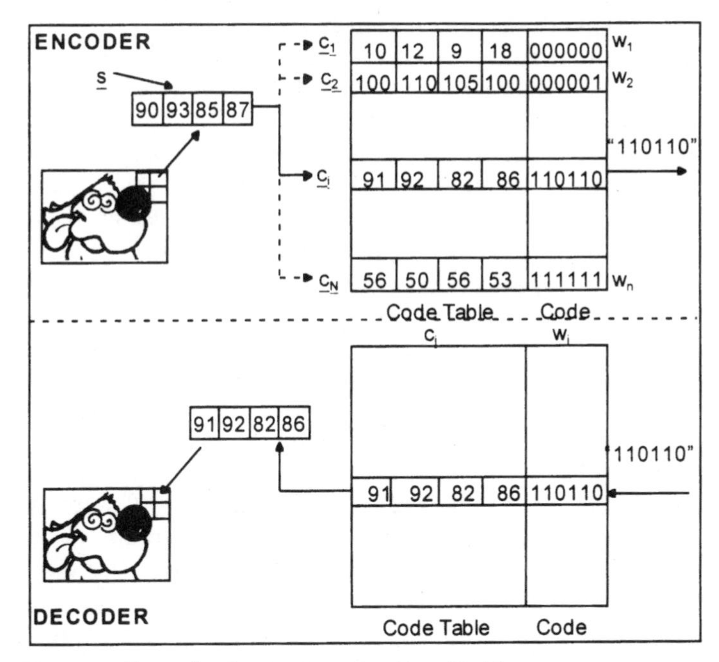

Figure 6.1 Example of vector quantization with $N = 4$.

which we call *code vectors.* Each code vector is coded with a distinct code word w_i. The table of all M code vectors is the *code book.* Finally, the distance $d\{\mathbf{c}_i, \mathbf{s}\}$ between code vectors and input vectors must be defined. As an error criterion d, we mostly use the quadratic equation:

$$d\{\mathbf{c}_i, \mathbf{s}\} = (s_1 - c_{i1})^2 + (s_2 - c_{i2})^2 + \ldots + (s_N - c_{iN})^2$$

This error criterion is not exactly optimum with respect to subjective perception, but at least it is the simplest to handle. Thus, all prior requirements for vector quantization have been met; now we will combine N number of values to be quantized $\{s_1,\ldots, s_N\}$, to give a vector $\mathbf{s}$. Then the vector $\mathbf{c}_i$ which is the most similar to vector $\mathbf{s}$ (and thus minimizes the distance $d\{\mathbf{c}_i, \mathbf{s}\}$) is retrieved from the code book. If this vector is found, its code word w_i is transmitted to the decoder. The decoder has access to the same code table and can, therefore, reconstruct the vector $\mathbf{c}_i$. This vector is inserted into the decoded picture. The quantizing error that occurs represents the distance $d\{\mathbf{c}_i, \mathbf{s}\}$.

We can look at the code book as a sort of building kit that contains a number of building blocks—the code vectors. It is the task of the vector quantizer to copy a given picture as closely as possible by using the available building blocks. The result will be sometimes better and sometimes worse, depending on the building kit and picture concerned. Actually, this method is known to us in connection with block matching (see Chap. 4 on motion-compensated interframe DPCM). A code book is also used for finding a code vector that is as similar as possible to an input vector, namely, the pattern block. The code book used for block matching comprises the blocks from the old picture that are located in the various search positions M. The search positions or motion vectors must likewise be coded with a code word. The only difference is that the code book used in block matching is different for every pattern block, while for VQ it remains unchanged after one initial design.

Each code vector $\mathbf{c}_i$ can be seen as a point in an N-dimensional room. If these points are arranged in a regular grid (or lattice), we can speak of a *lattice VQ*—a relative of linear quantization. If the code vectors do not form a regular structure, we speak of *random VQ*—a multidimensional version of nonlinear quantization. The many advantages and disadvantages of linear and nonlinear quantization are transmitted to their multidimensional versions. For the two-dimensional version ($N = 2$), Figs. 6.2 and 6.3 compare sections of a lattice VQ and a random VQ. An input vector $\mathbf{s}$ is allocated to the next code vector $\mathbf{c}$. The terms *Voronoi zone* and *neighbor* will be explained later. However obvious the idea of vector quantization might appear, the problems connected with it are equally manifold. How do we design a suitable code book that operates for all pictures? How do we

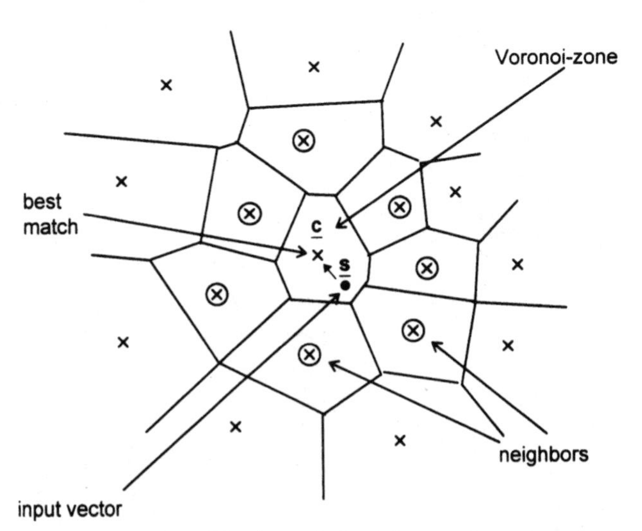

Figure 6.2 Random VQ; the code book has an irregular structure.

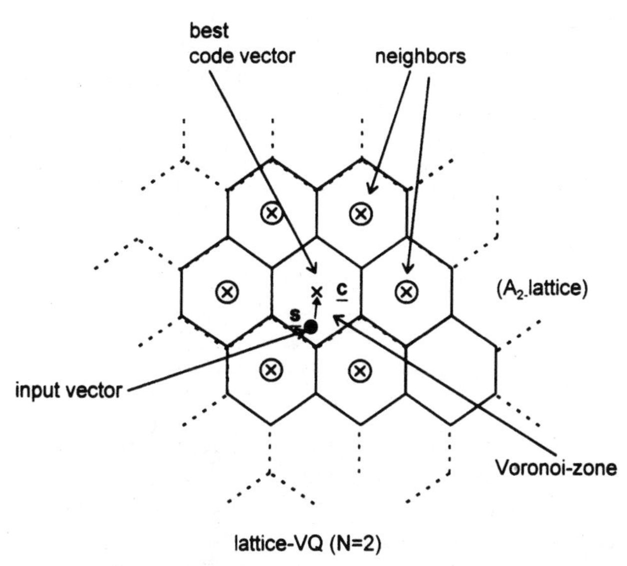

Figure 6.3 Lattice VQ; the code vectors are points in the regular A_2 lattice.

systematically find the best code vector for an input vector without having to try all the code vectors?

6.2.1 Random VQ: Designing the code book

The simplest and most popular method for designing random VQs, is, as in the case of scalars, the Lloyd algorithm, which was introduced in a previous chapter. Linde, Buzo, and Gray generalized it for vector quantization in 1980; hence, it is called the *LBG algorithm.*[1] In a practice session, we will see how typical picture contents are used to form a large number K of vectors

$$\boldsymbol{t}_i = \{t_{i1}, \cdots, t_{iN}\}$$

From these practice vectors, we acquire an initial code book in which the M code vectors $\mathbf{c}_1, \ldots, \mathbf{c}_m$, exactly M randomly selected vectors from the practice sequence, are read in. The LBG algorithm then operates as the Lloyd algorithm does:

1. Each practice vector is given exactly the code vector that lies next to it. Thereby a group of K_i practice vectors $\mathbf{t}_k(i)$ belongs to each code vector $\mathbf{c}_i$:

$$1 \leq k \leq K_i \qquad K_1 + K_2 + \cdots + K_M = K$$

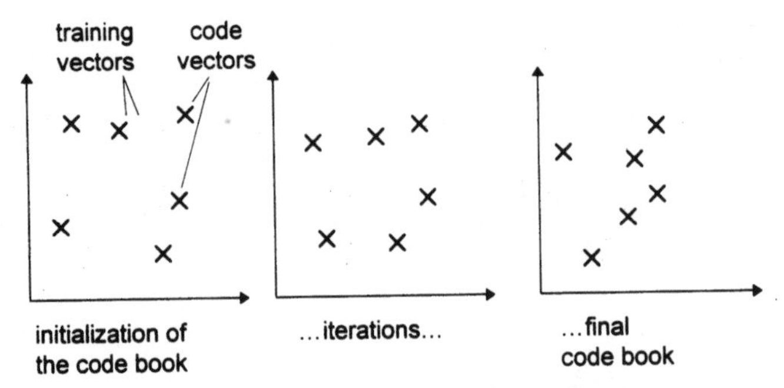

Figure 6.4 Operation of the LBG algorithm; code vectors move toward the training vectors.

2. Following this, each code vector $\mathbf{c}_i$ is displaced so that it represents the arithmetic average of the vectors $\mathbf{t}_k(i)$:

$$\boldsymbol{c}_i(\text{new}) = 1/K_i\ [\boldsymbol{t}_1(\text{i})+\boldsymbol{t}_2(\text{i}) + \cdots + \boldsymbol{t}_{Ki}(i)]$$

(vectors are added by adding the components). This new code book is used again in step 1.

The iterations 1 and 2 are alternated until the code vectors have sufficiently converged. The movement of the code vectors from iteration to iteration is shown in Fig. 6.4; the code vectors move preferably in regions in which the input vectors are concentrated. The form of these accumulations describes the memory of the input signal. Through the LBG algorithm, the code book can accommodate any memory, in contrast to transform coding, which is only suitable for linear statistically dependent correlations. This means, for example, in the two-dimensional case, that the input vectors occur mainly in the proximity of the angle bisectors $s_1 = s_2$ (see Chap. 5).

The convergence of the method and the final code book are greatly dependent on the choice of the initial code book. The random choice of practice vectors for forming the initial code book is not necessarily the best solution. Other methods of initialization are found in Ref. 1. With a small extension, the LBG algorithm can be used to design an *entropy-constrained VQ*.[2] The gain is not much greater than when the code vectors obtained with the simple LBG algorithm are Huffmann-coded (entropy-coded VQ).

The *pairwise nearest neighbor* (PNN) algorithm also operates with a practice sequence, but without an initial code book. Here, in each iteration, the two most similar vectors of the practice sequence are "melted" into one vector, so that the practice sequence is one vector smaller from iteration to iteration. The melting is nothing more than the formation of the weighted, arithmetic average of the vectors. The weight of such a vector is thereby proportional to the number of vectors from which it resulted by melting. If the practice sequence is shrunk to the desired number of vectors M, we can use it as a code book. In practice, the PNN algorithm requires too much calculation effort, generally speaking. There is, however, a very quick algorithm (the *Equitz algorithm* named after its discoverer[3]) which approximates a PNN algorithm. With the Equitz algorithm, the code book is generated approximately 20 times faster than with the LBG iteration and gives the same good result. The code book generated using the Equitz algorithm can be improved even further through some LBG iterations (a code book can only become better by applying an LBG algorithm, never worse).

A term which is always used in connection with vector quantizers is the *Voronoi zone.* This zone is the entry region of a code vector, within which all input vectors of this vector are quantized. In other words, an input vector $\mathbf{s}$ is assigned to the code vector $\mathbf{c}_i$ exactly when it lies in the Vornoi zone of $\mathbf{c}_i$. For the scalar quantizer ($N = 1$) this is the region between two decision thresholds, i.e., a straight piece within two points. In the two-dimensional plane, it is an area surrounded by several straight lines (Figs. 6.2 and 6.3). From three dimensions onward, the Voronoi zones are bordered by planes and hyperplanes. The neighbor of a code vector $\mathbf{c}_i$ is every vector $\mathbf{c}_i$ with which Vornoi zones of $\mathbf{c}_i$ have a common boundary.

Incidentally, the finding of a common neighbor in an unstructured code book is an unsolved mathematical problem whose solution would enormously reduce the calculation effort required for finding the code vectors. If a code vector is more similar to an input vector $\mathbf{s}$ than all its neighbors, it is also more similar than all remaining code vectors. It does not have to be compared any more. If we want to know the neighbors of all code vectors, we could systematically go from neighbor to neighbor in the direction of the best code vector.

6.2.2 Search strategies for the most similar code vector

The simplest method of finding the most similar code vector $\mathbf{c}_i$ to an input vector $\mathbf{s}$ is to draw all distances $d\{\mathbf{c}_i, \mathbf{s}\}$ where $1 \leq i \leq M$ and to select the vector with the smallest distance. Although this full search is a much used method, it requires heavy calculation. Help can be obtained by giving the code book a tree structure. This can lead to two different consequences, depending on the method.

1. The (binary) tree structure is generated during the code book design (*tree-searched VQ*[1]). When the VQ is carried out afterward, at M code vectors $\log_2 M$ distances $d\{*, *\}$ are calculated compared to M

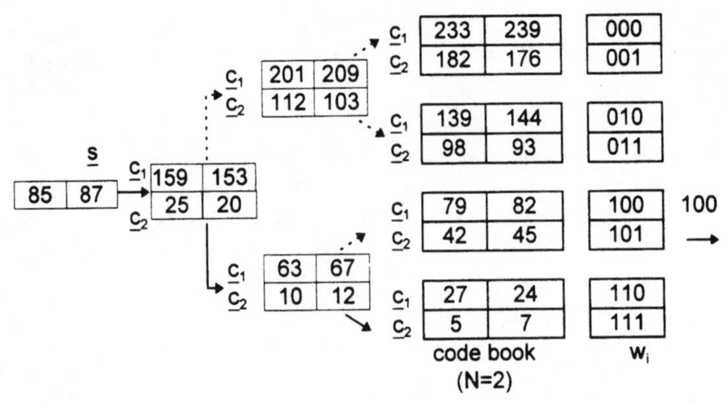

Figure 6.5 Tree-searched VQ; after three distance calculations the appropriate vector is found in the code book with $M = 8$ code vectors.

distances with the full search. The search method for $N = 2$ dimensions and $M = 8$ code vectors is shown in Fig. 6.5. In each of the three steps, a code book with only two code vectors is searched. We go either up or down, depending on which of these is the most similar of the two. Unfortunately, the code book of the VQ is not optimum nor is it certain that the most similar of all code vectors will be found. During the realization, it must also be taken into consideration that additional memory is required for the tree. Nevertheless, this form of VQ represents a good compromise between effort and compression gain, while for the same amount of calculation effort, a much larger code book can be used. The design method (*splitting-technique*[1]) for the tree and code book is also based on the LBG algorithm.

2. The code book is designed as usual and the tree structure is brought in afterwards. From here there are two approaches. The first works with so-called *k-d trees* and is explained in detail in Ref. 4. This algorithm finds the most similar code vector in a code book of M vectors, according to an average number of distance calculations proportional to log M. In the worst case, the calculation can take as long as that of a full search.

Another method tries to estimate the next neighbor of each vector by a gigantic calculation effort[5] and then to apply an algorithm which uses this knowledge in a tree structure.[6] Additionally, there are some mixed forms of search algorithms. There is still no algorithm that finds the best match and also consistently takes less time than the methods described previously. It is much simpler to find the most similar code word with the lattice VQ.

6.2.3 Lattice VQ

The code vectors in lattice VQ lie in a regular grid, and the shape of each Vornoi zone is identical. Every lattice can be described by a (nonsingular) matrix **L.** The multiplication of **L** with any vector $\mathbf{u}_i$ = $\{u_1,..., u_N\}$, whose components u_k are whole numbers, gives a valid lattice point and code vector

$$\boldsymbol{c}_i = \boldsymbol{L}\,\boldsymbol{u}_i$$

As we only want to accept a limited number $i \leq M$ of lattice points, the components of $\mathbf{u}_i$ move within given limits. The various lattices are labeled with a capital letter (for example, A_N, D_N, E_N, Z_N) whose index N gives the dimension. The simplest lattice is the Z_N lattice (**L** = unity matrix), whose lattice point, and thus whose code vectors, always have coordinates that are whole numbers. The next point to an input vector $\mathbf{s} = \{s_1, s_2\}$ in Z_2 is found by linear quantizing s_1 and s_2 independently with a step height of 1 (equivalent to a rounding off, Fig. 6.6*a*). For example:

$$\mathbf{s} = \{2.2, 3.7\} \qquad \mathbf{c}_i = \{2.0, 4.0\}$$

This lattice can be stretched, squeezed, and shifted by selecting other characteristics for the linear quantizing of s_1 and s_2 (Fig. 6.6*b*). The linear quantization can be interpreted as a lattice VQ.

The Z_N lattice can be rotated if $\mathbf{L} = \mathbf{A}^{-1}$ is selected, whereby **A** is an orthogonal matrix, as is also used for transform coding. The next lattice point in a turned lattice is found by transforming the vector **s**

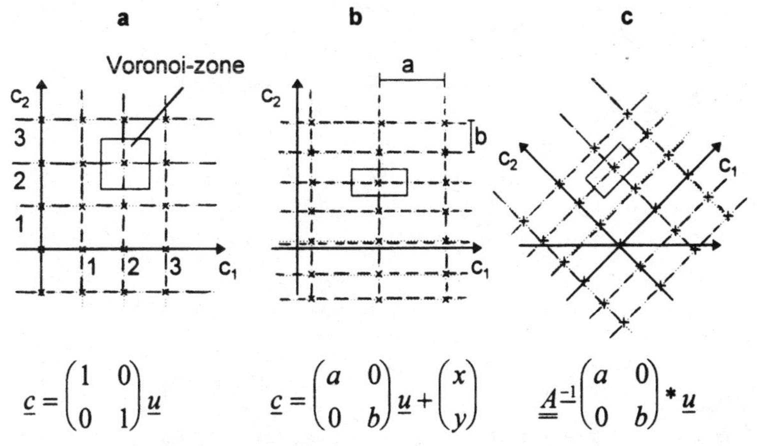

$$\underline{c} = \begin{pmatrix} 1 & 0 \\ 0 & 1 \end{pmatrix} \underline{u} \qquad \underline{c} = \begin{pmatrix} a & 0 \\ 0 & b \end{pmatrix} \underline{u} + \begin{pmatrix} x \\ y \end{pmatrix} \qquad \underline{\underline{A}}^{-1} \begin{pmatrix} a & 0 \\ 0 & b \end{pmatrix} * \underline{u}$$

Figure 6.6 Lattice VQ: (*a*) code vectors with grid points in a two-dimensional lattice, (*b*) scaling and shifting the grid by an alternative linear quantization, (*c*) lattice VQ with rotation of the grid as in transformation coding with linear quantization of the coefficients.

with **A,** and then linear quantizing the coefficients and retransforming the result (Fig. 6.6*c*). This method has already been discussed in Chap. 5: the transform coding with linear quantization of the coefficients can be understood as the elegant realization of a lattice VQ. The lattice is, in this case, a rotated and scaled version of the Z_N lattice. The code word w_i for a picture segment, is composed of the transform coding of the individual codes for the coefficients and the zero sequences. It could just as well be stored in a very long code-word table of given dimensions.

Only with the Z_N lattice do we find the lattice points by means of scalar quantization. A typical example is the so-called A_2 lattice (Fig. 6.3), which can be understood as a superimposition of two Z_2 lattices pushed together and stretched in the vertical direction by the factor [rd]3[xrd]; one of the lattices has its source in (0, 0), and the other in point (1/2, $\sqrt{3/2}$. The Vornoi zones then have a honeycomb shape. Here we look first of all (through the appropriate scalar quantization) in each of the two Z_2 lattices for the next point and select the better of these two candidates. The calculation effort is limited to two scalar quantizations plus two distance measurements, independent of how many code vectors there are. The search for the most similar code vectors in higher-dimensional lattices such as D_4 and E_8, for example, is explained in Ref. 7. Algorithms for any lattice can be found in Ref. 8.

At this point, an important characteristic of lattice VQs should be made clear: the most similar code vector is not the result of an exhaustive search but can in general be obtained with a few simple operations. With large code books, the calculation effort compared to random VQs is negligible.

6.2.4 Which lattice is the best?

On the one hand, we would like as large a piece of the N-dimensional space filled in with as few lattice points and Vornoi zones as possible, because in the end we have to spend a code word for every lattice point used. The Vornoi zones should, therefore, have as large a volume as possible. On the other hand, the distance of all possible input vectors of the next lattice points (that is, exactly the moment of inertia of the Vornoi zones) should, on average, be as small as possible. This is a well-known characteristic of the sphere: maximum volume with minimum moment of inertia. The more circular, spherical, or hyperspherical the Vornoi zones become, the better the quantizers become. Unfortunately, up to now this characteristic has only been possible for a few lattices, namely A_2, A_3, E_8, and the Leech lattice ($N = 24$). But we can also obtain useful results with other lattices.

The question still remains: How can we find the appropriate code word for a lattice point? With random VQ, the address of every code word can be stored together with the related code word. With lattice VQ, however, the code vectors are obtained; thus a code book is neither needed in the coder nor the decoder. One possibility is to portray the components $\{c_{i1},..., c_{iN}\}$ or $\{u_{i1},..., u_{iN}\}$ via an instruction $F\{*\}$ for consecutive whole numbers, i.e., a distinct numbering of the code vectors $\mathbf{c}_i$. The numbers can then be used as code word addresses, or, in their binary representation, as codes with a fixed word length. In exactly the same way, there must be an instruction $G\{*\}$ in the decoder to portray the code word and the number on the code vector. A simple example would be all u_i take values from 0 to 10. In this case, the components of u_i are taken as the decimal places of a decimal number which represents the desired number. Similar instructions can be traced from the characteristics of the lattice. The lattice points of the D_N lattice are thus so described that the sum of the coordinates c_{ij} of a code vector c_i is a figure that can be used for numbering. Alternative possibilities for coding lattice code vectors can be found in Ref. 8.

6.2.5 Working with less: Pyramid VQ

In 1988, Fisher made an interesting suggestion for drastically reducing the required code vectors and lattice points with the VQ of a memoryless source.[9] If we add the amounts s_i of N components of the input signal $\mathbf{s}$ together, we obtain for N approximately

$$s_1+s_2+\cdots+s_N = Nm$$

whereby m is the average of s_i. Clearly there are only certain input vectors $\mathbf{s}$ for N, namely, those for which the sum is the constant Nm. Why not, therefore, only allow such lattice points $\mathbf{c}_i$ of a lattice (e.g., Z_N) to be used as code vectors that fill the requirement

$$c_1+c_2+\cdots+c_N = Nm$$

Only one further selection of lattice points would be necessary for this, and therefore fewer code words are needed. Lattice points that meet these requirements are found (in the three-dimensional case) in a double pyramid ($c_1+c_2+c_3$ = constant), hence the name *pyramid VQ* (PVQ). This method works particularly well with Laplace-distributed sources, because in this case all code vectors from the PVQ are used, on average, with the same frequency. The code vectors can thus be coded with fixed code-word lengths, without efficiency loss. But the coefficients of subband coding are to a certain extent Laplace-distributed and memoryless. Therefore a PVQ works well here.[10] With the coding of such sources, the PVQ has the advantage over the simple lattice quantization because these only work efficiently with entropy coding.

The PVQ needs about $4N$ multiplications to quantize an input vector. Fisher was able to show that the efficiency of a PVQ can be compared approximately with that of a scalar, linear quantizer, if coding was with a variable word length. The gain also lies in the saving of the data buffer. Pyramid PVQ should not be confused with *pyramid coding,* which is a form of subband coding.

6.2.6 Lattice or random VQ?

It would appear that random VQ is superior to lattice VQ. Lattice VQ is only optimum for cases of memoryless, evenly distributed input vectors. Consider once more Fig. 6.4 in relation to this: no regular lattice could perfectly fit the indicated frequency distribution of the input vectors, based on the same number of code vectors. The same problem occurs with the linear quantizer in the scalar case: linear quantization is easy to realize because of weighting and rounding but is inefficient if the input signal is not evenly distributed. Despite this, if we use a variable word length for coding the substitute values (ECQ), the efficiency will increase considerably. Furthermore: the optimum ECQ for many substitute values is a linear quantizer, independent of the distribution of the input signal. A similar situation applies for vector quantization: the optimum vector quantizer for a large number M of code vectors, if the code vectors are to be entropy-coded (e.g., Huffmann code[11]), is a lattice vector quantizer. The number M can in fact be very large, in the odd case.

Furthermore, we can raise the efficiency of lattice VQs by not using the same lattice point $\mathbf{c}_i$ as the code vector at the decoder, but another point $\mathbf{c}'_i$ in the same Voronoi zone, which generates the smallest error in relation to a typical practice sequence.[12] This point does not necessarily lie in the center of Vornoi zones but is calculated from the average of the practice vectors in the respective Vornoi zones (Fig. 6.7). To find these points $\mathbf{c}'_i$ for all the lattice points $\mathbf{c}_i$, in principle we carry out (once) steps 1 and 2 of the LBG algorithm. Further iterations of the LBG algorithm would further increase the coding efficiency, but then the structure of the lattice VQ would be lost and the best match would have to be searched for again. With this method, a code book must at least be available at the decoder.

Finally, there is the possibility, as with linear quantization, to predistort the input signal with a nonlinear characteristic applied prior to the lattice VQ. As with the random VQ, we can incorporate the statistical properties of the input signal with all the implementation

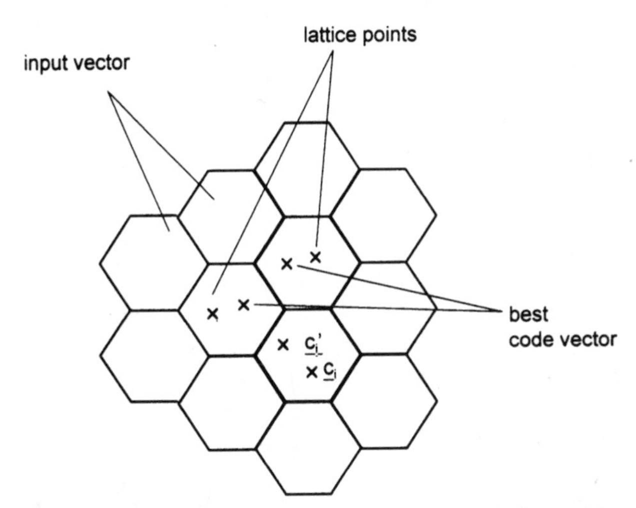

Figure 6.7 The best code vector is not necessarily located in the center of the Voronoi zone.

advantages of the lattice structure. The signal must next be equalized, again with the inverse characteristic. The simplest possibility is to compand each vector component one-dimensionally. For an optimum result, especially with signals with memory content, real multidimensional companding must take place. The multidimensional companding characteristic must also be optimized for the picture contents in question. This procedure has not yet been researched extensively, and it has been shown that the multidimensional companding is not possible with some signal sources.[11] A high-quality DPCM picture coding (4 bits/pixel, approximately 50-dB PSNR), for which the prediction error is subjected to a lattice VQ, after a one-dimensional companding, is described in Ref. 22. A further important characteristic of lattice VQ is that the maximum quantization error represents exactly one half of the diameter of the always identical Vornoi zone (companding, overloading of the VQ is for once excluded). With random VQ, however, all Vornoi zones have different sizes, and individual Vornoi zones can therefore be very large. Consequently, with random VQ, we can never prevent the occurrence of very strong local coding errors, which hardly influence the average coding error but are subjectively disturbing. It is, therefore, difficult with an ordinary random VQ to guarantee really good picture quality (even with high data rates), let alone solve the problem of designing and storing the necessary code books in their required sizes.

Experiments up to now have been carried out in most cases with random VQs instead of lattice VQs, even if the latter has become more popular in recent times. The following reasons may account for this:

1. Especially with input signals with memory content, input vectors concentrate at specific accumulation points of multidimensional space, as already indicated in Fig. 6.4. If we do not expect too much of the coding result, such signals can be coded with comparatively few vectors at a low data rate (e.g., 1:40, 1:80). In that case, entropy coding would be used at the very least. That is the actual strength of the random VQ: medium quality at a small data rate and a fixed word length. For a reasonable vector dimension N—typical values are $4 \leq N \leq 25$—we need on the order of $200 \leq M \leq 2000$ code vectors with random VQ. To achieve comparable compression results with a lattice quantizer and entropy coding, the number of lattice points and necessary code words are an order of magnitude higher. Storage of the code books can be avoided with lattice VQ (but not the storage of the code words) if coding is with a variable word length. Additionally, there are the typical problems with entropy coding with respect to synchronization, buffering, and error protection. Lattice VQ is preferred for signals with a small memory depth and a small dynamic range, because with a small dimension N, few lattice points are required.
2. Although the calculation effort of the lattice quantizer is small, the algorithms are still frequently tricky, which can cause problems for the numbering of the lattice points and the quantizing of vectors that lie outside of the dynamic range. Software implementation of the LBG algorithm and a full-search random VQ is, however, very simple.

A typical disadvantage of random VQs, one that has only recently become apparent, is the so-called *inside-outside effect.* If we use the same code book on picture contents that has been used for practice, the compression results are much better than for contents that have not been used for practice. The latter is the normal case, as we do not want to design and transmit a new code book for each picture. We can control this problem to a certain degree by adapting the code books to the momentary picture content,[24,25] but only at the cost of a disproportionately high calculation effort. There are other ways of reducing the inside-outside effect, which are artificially generated by the practice sequences according to universal, statistical picture models;[13] however, real compression results from this method are unknown to the author.

A general judgment on the best VQ method is impossible. As always, it depends on finding the best method for the given application. In the next section we will sketch how the various VQs can best be used.

6.3 VQ Applied to Picture Coding

The direct application of VQ to pictures as shown in Fig. 6.1 makes little sense, generally speaking, because neighboring relations within natural pictures are too large. To take advantage of this to some extent, block sizes of approximately 8×8 pixels would be required, which equates to a vector dimension of $N = 64$. The number of code words required for an acceptable search (also for random VQ) would be too high and for a full search, too exhaustive. Also the usual error criterion $d\{*, *\}$ for the subjective perception is not really suitable. The following three solutions can be considered, both by themselves and in combination.

6.3.1 Classified VQ (CVQ)[14,15]

With coding, it is more important to ensure that the assigned code vector receives (qualitatively) the same picture structure than it is to find the code vector with the smallest distance. It is also important, for the subjective feeling, that the picture edges are reproduced correctly with respect to direction and contrast. A promising approach for good subjective results is, therefore, to design several code books for the various types of structures—for example, a code book for vertical edges or one for regions with less detail (Fig. 6.8). By applying several code books, the full search within all code vectors would be avoided. The use of detail classes is better for the subjective picture quality than the blind application of an insensitive full search.

The coding sequence proceeds as follows: the input vector is examined and assigned to one of the classes. Then it is vector-quantized with the appropriate code book, and coded. A block size no larger than

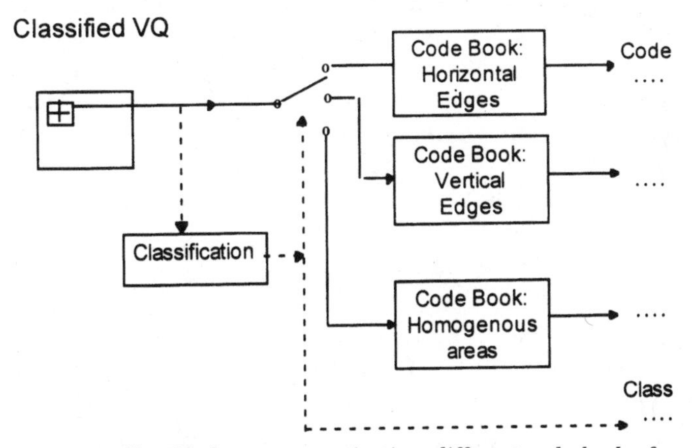

Figure 6.8 Classified vector quantization: different code books for different structures.

5×5 pixels should be chosen, to avoid the requirement to take into account curved edges, or the frequent occurrence of two edges in one block. The simple CVQ uses memory only partly for this. The decoder, in addition to the code for the code vector, must also be informed of the class in coded form. With this coding technique, coding results have been obtained that perform equally to some transform coding schemes.

6.3.2 Applications with high-pass and bandpass signals

The block size, and hence the number of necessary code vectors, can be reduced by vector quantizing not the entire picture signal but only the high-frequency parts. The low-pass part is coded separately (e.g., PCM, DPCM, or TC) and transmitted separately at a low data rate. Because the low-pass component contains the large memory of the picture signal, there is a much smaller memory with the high-pass part. Here, lattice VQ lends itself well, as does random VQ. Because picture objects stay qualitatively intact after high-pass filtering, we can alternatively design a classified VQ for the high-pass part.

Here are some examples:

- We subtract from each input vector its mean value and use the VQ on the remainder. Additionally, the mean value must be quantized and coded (called *separating-mean VQ*[1]).
- After prefiltering, we subsample the picture by an order of 4 or 8 and transmit it to the decoder with a small data rate. Through interpolation of the subsampled picture, the low-pass component is obtained in the encoder and decoder (Fig. 6.9). It is subtracted from the original in the encoder. The difference frame is then segment-wise vector-quantized, transmitted, and added to the low-pass part at the decoder (called *interpolative VQ*[16]). This low-pass component well-represents the picture content in homogeneous surfaces. Thus the VQ can concentrate on coding the detailed part of the picture. The blocking effect, which is particularly disturbing in picture areas without detail, is avoided to some extent with this approach.
- We use VQ on the high-frequency coefficients by transform coding (*TC VQ*[8]), or on the high-frequency subbands after the subband splitting.[17] Contrasting forms of VQ can be found in Refs. 10 and 12. The tendency of the eye to be insensitive toward high-frequency coding errors can be exploited by coding the higher frequencies with increasingly fewer bits, thus saving bits.
- Also, the subtraction of two successive pictures can be interpreted as a high-pass filtering. That is why difference frames are also

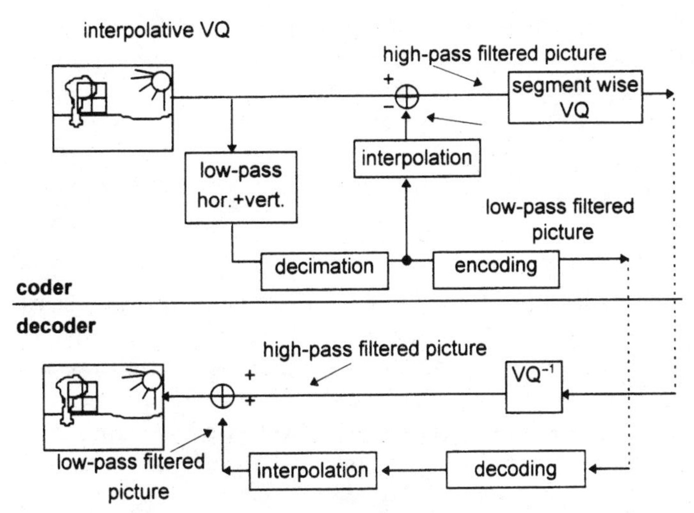

Figure 6.9 Interpolative VQ: only the high frequencies of the picture signal are vector-quantized.

suitable for VQ. An interesting approach can be found in Ref. 18. Here, the input vector is being vector-quantized in a first step, with a (tree-searched) random VQ. The difference of the input vector and the code vector is subsequently subjected to a lattice VQ. A code word must, of course, be transmitted for both VQs.

6.3.3 Previsualized VQ

The picture is subjected to reversible preprocessing, which corresponds to a rough approximation of the subjective human perception.[23] Following this, the preprocessed signal is vector-quantized and vector-coded. The decoder reconstructs the code vector and reverses the preprocessing. Through this preprocessing, which comprises nonlinear amplitude distortion and filtering, the objective error dimension $d\{*, *\}$ is converted to an error dimension that is better adapted to the subjective perception.

As always with picture coding, there are no limits to what one can imagine, and the same goes for VQ. Therefore, in Ref. 19, for example, a combination of interpolative VQ, classified VQ, and TC VQ is suggested. A selection of different techniques can be found in Ref. 20. Typical for most of these VQ coders is a picture quality that is still acceptable even at high compression factors (1:40 to 1:80 with intracoding). Somewhat more difficult is high-quality coding at high bit rates, as there is an explosion in the number of code vectors necessary.

We assume that, especially with high-pass signals and difference pictures, statistical dependencies of a higher order of magnitude exist, which can be exploited better by VQ than by the remaining decorrelation methods, such as TC or DPCM.

6.3.4 Vector quantization and error protection

The central idea of seeing a code vector as a point in an N-dimensional space, which has a certain distance to other code vectors, is the basis of *channel coding*. That is why vector quantizing and transform quantizing are strongly related to channel coding. The following text is only intended to sketch the relationship between channel and source coding and give a rough feeling for how and why channel coding works, not how it is realized. The technology of VQ, introduced previously, is further applied in the following, although the usual terminology does not necessarily match exactly that of channel coding. For an introduction to this area, a good textbook is Ref. 21.

Leaving aside the implementation issues, we can also use a vector quantizer for channel coding instead of for compression, by simply exchanging the encoder and the decoder: we then use the actual decoder as channel encoder and the encoder as channel decoder. The coding procedure functions as follows (Fig. 6.10): a code word w_i—8 bits long, for example—has to be transmitted. A code vector $\mathbf{c}_i$ from the code book, belongs to each code word w_i. This code vector is digital-to-analog converted and sent via the channel. It is thereby superimposed with a disturbance $\mathbf{s}$. It is analog-to-digital converted again at the decoder. Then the most similar code vector $\mathbf{c}_i$ to the received vector $\mathbf{c} + \mathbf{s}$ is searched for in the code book. A distinct code word w_j belongs once more to the code vector $\mathbf{c}_j$. This code word corresponds exactly to the code word transmitted, provided the disturbance was not too large, or more precisely, if the sum $\mathbf{c}_j + \mathbf{s}$ is still in the Vornoi zone of $\mathbf{c}_j$.

According to which criteria do we optimize a code book for channel coding? Three points (for which the first two also apply to lattice quantizers) must be taken into consideration here:

- The dispersion of output energy should be restricted as much as possible. Therefore, the code vectors should be packed as closely together as possible.
- On the other hand, there should be as much distance between the code vectors as possible. But a note of caution: depending on the channel characteristics, the code vectors will be affected differently, which leads us to the third condition.

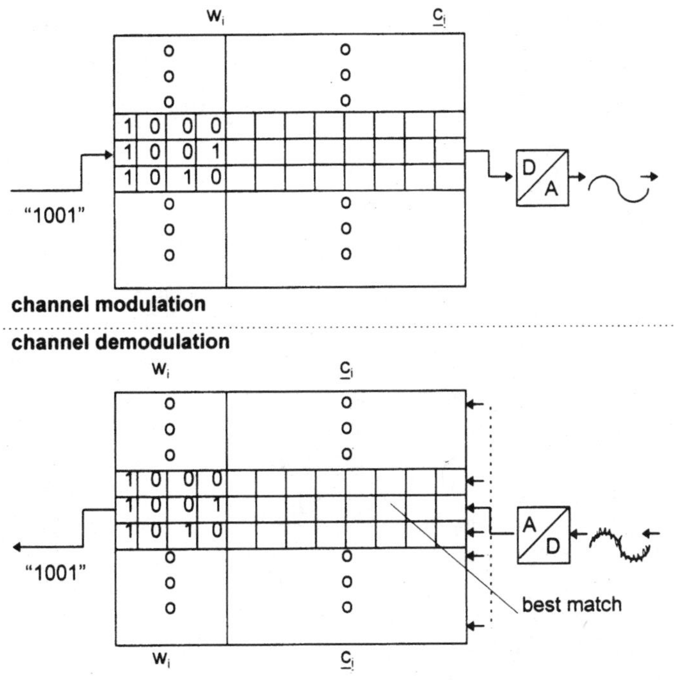

Figure 6.10 Exchange of the encoder and decoder; the vector quantizer becomes a channel coder.

- The signal should be as well-suited to the channel as possible, with consideration for bandwidth, transmission characteristic, dynamic range, etc.

This form of channel coding offers the best error protection for the highest possible number of code vectors, which will then be correspondingly long. Because of the amount of effort that would otherwise be involved, no code book is used for the coder and decoder in practice. Instead, the vectors are generated with modulation techniques (PSK, FSK, 16 QAM, etc.). If, for example, the code vectors represent sine-wave–shaped signals with different phases, this method corresponds with a frequency and phase modulation.

Alternatively, we can generate code vectors with a transform. As already explained, this gives us code vectors on a rotated Z_N lattice; the binary code word is made available at the input of a transform. The transform of this binary vector corresponds to the code vector to be transmitted $\mathbf{c}_i$. If we use an inverse FFT for the transform, the output signal is the superimposition of N sine waves of different frequencies. This gives us a wideband signal. To decode, we can carry out an FFT on the received signal and round off each output of the FFT to 0

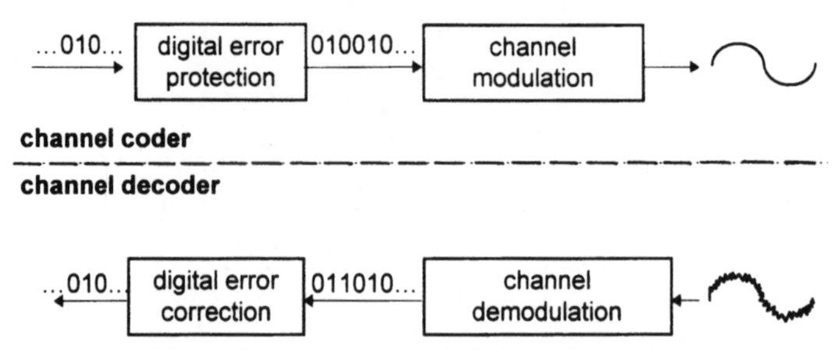

Figure 6.11 Channel coding composed of channel modulation and digital error protection.

and 1. This form of error protection (called *soft decision*) is labor-intensive for large dimensions of N. Therefore, it is embedded in a suboptimum, but—for large vector dimensions—realizable, digital error protection (Fig. 6.11).

Also, simpler digital *hard-decision* error protection could be realized with VQ. In the code vectors $\mathbf{c}_i$, only zeros and ones occur. If we calculate the distance $d\{\mathbf{c}_i, \mathbf{s}\}$ between the vectors $\mathbf{c}_i$ and $\mathbf{s}$, we get the same result as if we had simply counted the number of different binary places between both vectors. For example:

$$\mathbf{c}_i = \{1, 1, 0, 0, 1\}$$

$$\mathbf{s} = \{0, 1, 0, 1, 1\} \quad d\{\mathbf{c}_i, \mathbf{s}\} = 2$$

This distance is called the *Hamming distance*. The coding procedure runs exactly as in the previous case (Fig. 6.12): a code book contains a distinct yet binary code vector $\mathbf{c}_i$ for each code word w_i. The code vector $\mathbf{c}_i$ is transmitted. On the channel, $\mathbf{c}_i$ is distorted, i.e., one or more bit places are flipped over. A code word $\mathbf{c}_k$ that is as similar as possible to the received code vector is searched for in a code book at the decoder. If $\mathbf{c}_i$ is not too strongly distorted, then $\mathbf{c}_k = \mathbf{c}_i$, and the decoder gives the right code word w_i. If two code words are equally similar, it is at least known that an error occurred.

As we can see from Fig. 6.12, we pay for the digital error protection in that the code vectors $\mathbf{c}_i$ are some bits (called *protection bits*) longer than w_i, i.e., the word to be transmitted. In this crude system, we have increased the data rate by 200 percent. The art is to find a code book that needs as few protection bits as possible and has the largest Hamming distance $d\{*, *\}$ between two code vectors (pair-wise).

More appropriately, this coding is not carried out with tables but uses strong structured code books. The code vector $\mathbf{c}_i$ can then be calculated from the code word w_i and vice versa (see lattice VQ).

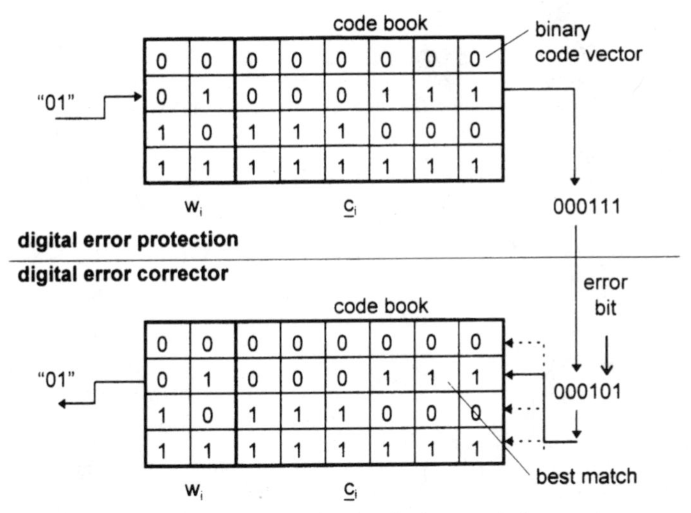

Figure 6.12 Digital error protection can be interpreted as vector quantization.

Although we are reminded of the lattice VQ here, these binary structures have no longer any geometric interpretation, as for example a lattice, but are of an algebraic nature. The Reed-Solomon code is widely in use.

Error protection, while commonly implemented, is mostly beyond the scope of this book on digital video and audio compression. Discussion of error detection, correction, and protection is more appropriate in discussions of communication systems. However, some discussion of the topic is unavoidable when evaluating compression schemes, because a bad bit may be disastrous in some approaches while hardly noticeable in others.

Vector quantization is a comparatively young technique and this approach is, again, only a momentary picture of the state of the art. Many limitations of VQ are due not so much to their principal nature, but instead to dependency on the current calculation and storage capacity available, a problem which is reminiscent of the state of transform coding only a few years ago. It is to be assumed, therefore, that the importance of VQ will increase in the next few years.

While vector quantization has not succeeded (to date) in taking the step from the laboratory to practical application, another method probably has better chances of being used (in certain areas of picture transmission) in the near-term: *subband coding*. This coding technique, which is also related to transform coding, will be the subject of the next chapter.

6.4 References

1. Gray, R. M., "Vector Quantization," *IEEE ASSP Magazine,* April 1984, p. 4–28.
2. Chou, P. A. M., T. Lookabaugh, and R. Gray, "Entropy Constrained Vector Quantization," *IEEE Trans. ASSP,* vol. 37, January 1989, pp. 31–42.
3. Equitz, W., "A New Vector Quantization Clustering Algorithm," *IEEE Trans. ASSP,* vol. 37, October 1989, pp. 1568–1575.
4. Friedman, J. H., J. L. Bentley, and R. A. Finkel, "An Algorithm for Finding Best Matches in Logarithmic Expected Time," *ACM Trans. Math. Software,* vol. 3, September 1977, pp. 209–226.
5. Gersho, A., "On the Structure of Vector Quantizers," *IEEE Trans. Inform. Theory.,* vol. 28, March 1982, pp. 157–166.
6. Gersho, A., and D. Cheng, "Fast Nearest Neighbor Search for Nonstructured Euclidean Codes," *Abstracts of the 1983 IEEE Intern. Symp. on Theory,* September 1983, p. 88.
7. Conway, J. H., and N. J. A. Sloane, "Fast Quantizing and Decoding Algorithms for Lattice Quantizers and Codes," *IEEE Trans. Inform. Theory,* vol. 28, March 1982, pp. 227–232.
8. Saywood, K., J. D. Gibson, and M. C. Rost, "An Algorithm for Uniform Vector Quantizer Design," *IEEE Trans. Inform. Theory,* vol. 30, November 1984, pp. 805–814.
9. Fisher, T. R., "A Pyramid Vector Quantizer," *IEEE Trans. Inform. Theory,* vol. 32, July 1986, pp. 568–583.
10. Blain, M. E., and T. R. Fisher, "A Comparison of Vector Quantization Techniques in Transform and Subband Coding of Imagery," *EURASIP Image Communication,* vol. 3, February 1991, pp. 91–105.
11. Gersho, A., "Asymptotically Optimal Block Quantization," *IEEE Trans. Inform. Theory,* vol. 25, July 1979, pp. 373–380.
12. Girod, B., and T. Senoo, "Entropy-Coded Vector Quantization of Image Subbands," *Session 4, Dortmund TV Seminar,* October 1991, pp. 124–132.
13. Du, Y., and H. Docter, "Model-based Generation of Pseudo-picture Blocks for Practice Vector Quantization," 7. *Aachen Symp. for Signal Theory,* Ameling, Springer, 1990, pp. 90–95.
14. Ramamurthi, B., and A. Gersho, "Classified Vector Quantization of Images," *IEEE Trans. Comm.,* vol. 34, November 1989, pp. 1105–1115.
15. Kim, D. S., and S. U. Lee, "Image Vector Quantizer Based on a Classification in the DCT Domain," *IEEE Trans. Comm.,* vol. 39, April 1991, pp. 549–556.
16. Hang, H. M., and B. G. Haskell, "Interpolative Vector Quantization of Color Images," *IEEE Trans. Comm.,* vol. 36, April 1989, pp. 465–470.
17. Ho, Y. S., and A. Gersho, "Classified Transform Coding of Images Using VQ," *IEEE-ICASSP,* Glasgow, 1989, pp. 1890–1893.
18. Bage, M. J., "Interframe Predictive Coding of Images Using Hybrid Vector Quantization," *IEEE Trans. Comm.,* vol. 34, April 1989, pp. 411–415.
19. Westerink, P. H., *Subband Coding of Images,* Dissertation TU Delft, Holland, 1989.
20. Nasrabadi, N. M., and R. A. King, "Image Coding Using VQ: A Review," *IEEE Trans. Comm.,* vol. 36, August 1989, pp. 957–971.
21. Chambers, W. G., *Basics of Communications and Coding,* Oxford Science Publications.
22. Chen, T. C., "A Lattice VQ Using a Geometric Decomposition," *IEEE Trans. Comm.,* vol. 38, May 1990, pp. 704–714.
23. Xie, Z., and T. G. Stockham, "Previsualized Image VQ with Optimized Pre and Postprocessors," *IEEE Trans. Comm.,* vol. 39, November 1991, pp. 1662–1671.
24. Goldberg, M., and H. Sun, "Frame Adaptive VQ for Image Sequence Coding," *IEEE Trans. Comm.,* vol. 36, May 1988, pp. 629–635.
25. Panchanathan, S., and M. Goldberg, "Adaptive Algorithms for Image Coding Using VQ," *Image Communication* (Eurasip publ.), vol. 4, no. 1, November 1991, pp. 89–92.

Chapter 7

Subband Coding

7.1 From Sound to Vision

Subband coding (SBC) is commonly used for audio coding. Imagine that a piece of music for double bass and piccolo—two instruments with completely different tones—must be transmitted with as few bits as possible. This has a direct effect on the spectrum of the piece of music: it would be possible to ascertain that there is an average frequency band in which no output is generated. The question follows: Why not filter out the bass region and the high notes, code them separately, and eliminate transmission of the middle bandwidth? Although this example greatly simplifies the problem represented, it identifies the core of the matter: subband coding is based on the uneven distribution of the signal level over the entire frequency spectrum.

For high-quality audio transmission, a data compression factor of approximately 4 to 8 can be achieved with SBC without noticeable quality loss. Quantization noise, caused by the quantizing of a subband, is limited to the appropriate subband. The SBC can therefore take advantage of the perception properties of the ear, especially masking effects in the frequency domain, along with the signal properties. Furthermore, we can quantize uncritical subbands more coarsely, because of the frequency-dependent sensitivity of the ear (*noise shaping*).

The use of SBC for pictures came later, because masking effects in the frequency domain are not useful where the eye is concerned. Nevertheless, the eye also perceives higher local frequencies less well, so that in the higher-frequency subbands, larger errors are permissible. Alongside these characteristics of the visual system, the redundancies of the picture signal are exploited very well by SBC. Furthermore, there are practical advantages: low-frequency subbands lend themselves to the generation of a low-resolution picture without much effort. This is good for scaled transmission, error protection, and progressive transmission, as will be explained later.

7.2 Principles of Decorrelation

The first step in coding is frequently to decorrelate the picture signal, i.e., try to extract from the signal the linear statistical dependency.. The memoryless signal can be subsequently scalar quantized and coded, without a large loss in efficiency. With DPCM, the decorrelation occurs through filtering. The resulting predicted error signal shows only a small output. With transform coding, it can be determined that the coefficients have almost no linear dependencies anymore. On the other hand, the output is concentrated on only a few coefficients.

What does this depend on? The source model supplies an answer as it comprises an internal dc-voltage–free, white noise source, followed by a linear model filter. The linear filter ensures a picture signal with a typical power density spectrum in which the lower-frequency bands contain the main component of the output signal.

The SBC is put to work here: we split the signal into its spectral components with several bandpass filters. The outputs of two bandpass signals of a linear process are not correlated if the bandpasses do not overlap. The bandpass parts of the signal indicate no linear dependency among each other; thus each bandpass signal can be quantized and coded as a scalar. If the subbands are narrow enough, the output power spectrum inside a band is also relatively flat and is, therefore, also a subband, which points to a decorrelation of the sampled values within the subband. The effort is repaid here because the output in some subbands is often very small, especially where the model filter of the source model shows a large attenuation. With natural pictures, a concentration of output can be seen at the low-frequency subbands. The usually low output power of the remaining subbands produces a small bit rate. The similarity with transform coding is obvious. We will go into this in more detail in the next section.

7.2.1 Basic diagram

It is assumed that the successively read out lines of a picture are to be coded because this is the only way to use the horizontally oriented statistical dependencies. First of all, we will examine a provisional, idealized diagram of subband coding, as shown in Fig. 7.1. The signal is split into M subbands via nonoverlapping bandpass filters. For the sake of simplicity, we will assume that the bandwidth of all filters is the same. Unfortunately, after the band splitting, instead of one pic-

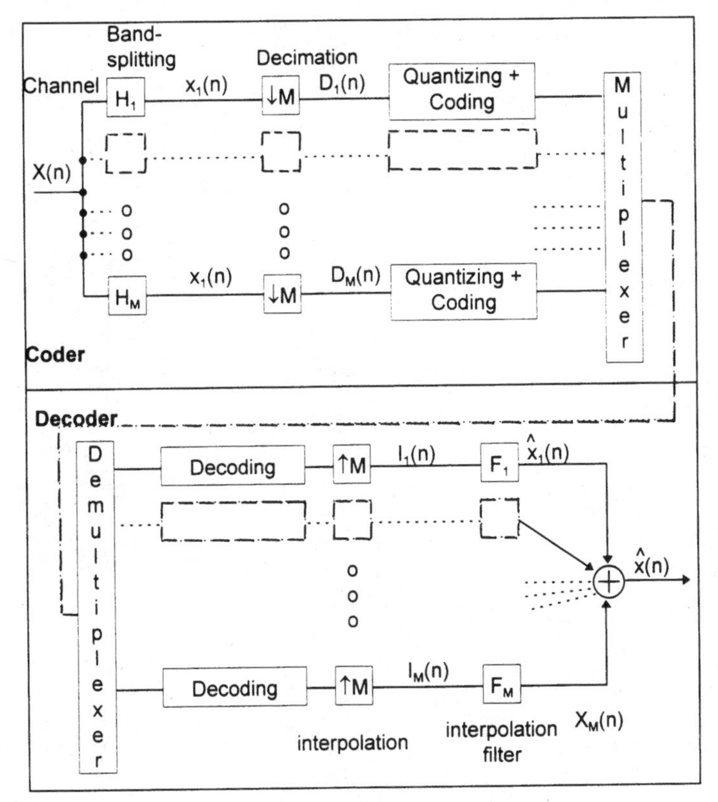

Figure 7.1 System for subband coding with scalar quantization of the subbands.

ture with pixels×lines samples, we suddenly have M pictures with pixels×lines samples: a data expansion by factor M. A decimation of the amount of samples is required.

This procedure is shown for $M = 2$ in Fig. 7.2*a* through *f*. Because of sampling, all spectra are repeated many times around multiples of the sample frequency at $\omega = 2\pi$. In the first step, the band splitting in $X_1(\omega)$ and $X_2(\omega)$ occurs via ideal, nonoverlapping filters (Fig. 7.2*b*). With the subsequent decimation, each second value is set to zero in each subband. This can be achieved by multiplying the succession of samples $X_1(n)$ and $X_2(n)$ with sequence $0.5 + 0.5 \cos n\pi = \{1, 0, 1, 0, 1,\ldots\}$. Via this obvious modulation procedure, a second spectrum is made that fits exactly in the gaps of the low-pass signal.

With the high-pass branch, the gaps in the lower-frequency region are filled with the spectrum. If we leave out the zeros from the transmission, the information content will not change, but the signals are compressed somewhat in the time domain. This shows up as a rescal-

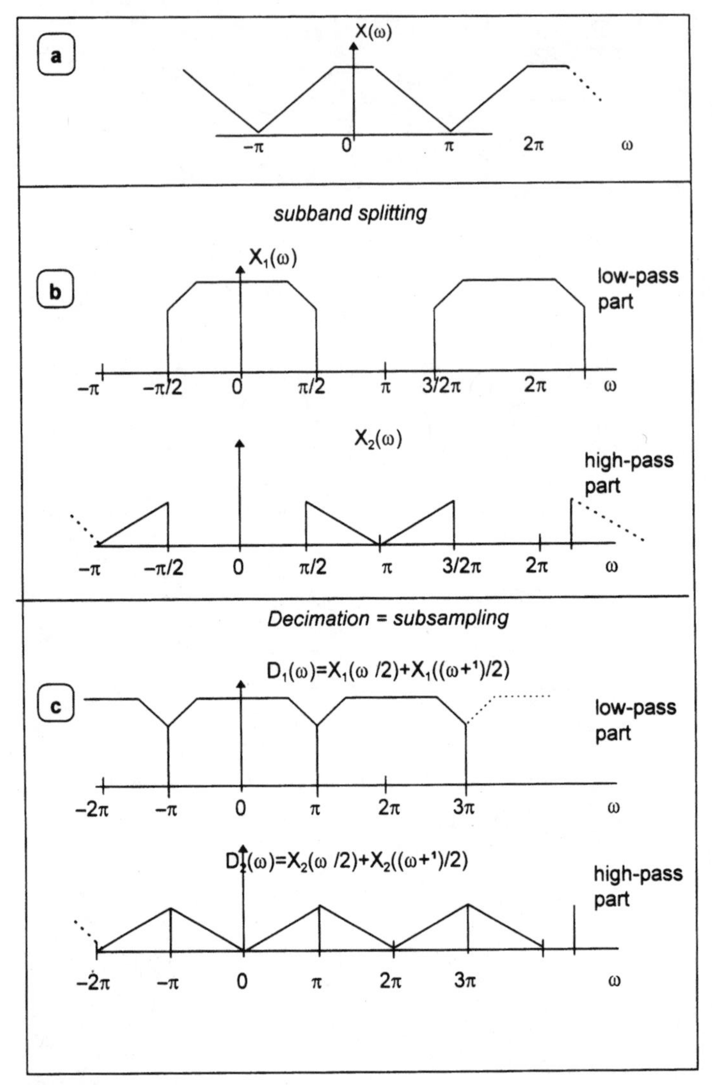

Figure 7.2 Subsampling and interpolation with nonoverlapping band-pass filters.

ing of the ω axis in Fig. 7.2*c*. If we reinsert the zeros during interpolation, after transmission, the signal paths open up again for a while so that the original scaling of the ω axis is restored (Fig. 7.2*d*). The parts of the spectra that were added during decimation can now be removed by the interpolation filter at the decoder (Fig. 7.2*e*). If we now add both signals, we will get the original signal, although before samples in the subbands were set to zero (Fig. 7.2*f*).

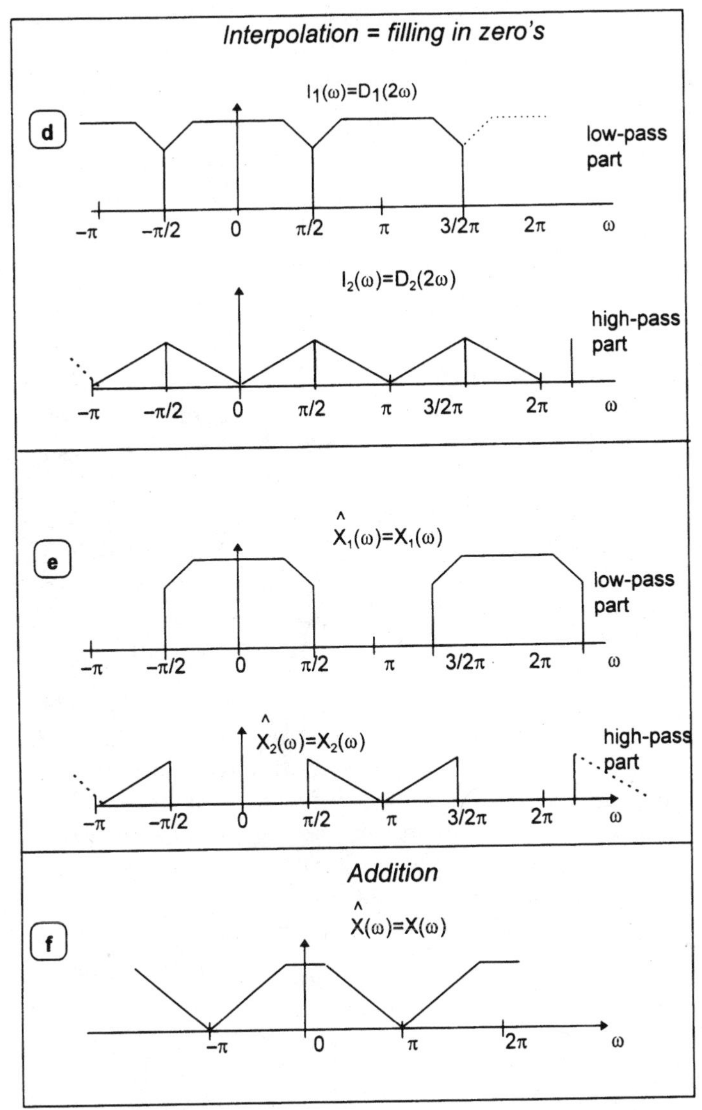

Figure 7.2 (*Cont.*)

The same goes for splitting the signal into M subbands. With ideally band-limited subbands, only each Mth sample is to be transmitted. The interpolation at the decoder inserts a zero between each of the samples M, after which the respective subband is filtered out again. The sum of the subband signals again returns the original signal.

The samples must be quantized before actual transmission. In the simplest case, this can be done by scalar quantization of the samples of the subbands. But for now the system is complete: band splitting, decimation, scalar quantizing, encoding, decoding with reconstruction of the quantizer substitute values, interpolation, interpolation filtering, and addition of the subbands—these processes are the components of subband coding.

7.2.2 Gain

What do we gain from subband coding compared to direct quantization and transmission of the signal (PCM)? There are two equivalent possibilities for expressing the gain:

1. In Chap. 5, for example, the data rate of transform coding and PCM were compared, if both were allowed to generate the same coding error in the form of a mean squared deviation. This method would also be applicable for subband coding.
2. The alternative way is as follows. The same data rate is given for PCM and subband coding, and then we ask how much smaller the coding error of subband coding is, compared with that of PCM. The way in which the available data rate is divided among the subbands, in order to maximize the coding gain, is also determined.[8]

First of all, we must determine how bad the PCM actually is by obtaining the coding error output $\sigma_{q,\mathrm{PCM}}^2$ by directly quantizing the signal. A rule of thumb for this was given in Chap. 3, by which, at least with high bit rates, the error output can be estimated, depending on the input power σ_x^2 and the PCM bit rate:

$$\sigma_{q,\mathrm{PCM}}^2 = c \times 2^{-2} \times R_{\mathrm{PCM}} \times \sigma_x^2$$

Band splitting causes the output of the input signal σ_x^2 to be distributed over the subbands, so that the sum of the individual outputs σ_{xk}^2 again produces the combined output:

$$\sigma_{q,\mathrm{SBC}}^2 = \sum_{k=1}^{m} \sigma_{xk}^2$$

If ideal, nonoverlapping filters are used for subband coding, an error output of σ_{qk}^2 will be generated in each of the subbands $k = 1,\dots, M$, by the quantization, which is added to each respective subband signal. This error output is a result (as above) of the bit rate for the subband R_k and the output of the subband σ_{xk}^2. The combined error output being searched for, i.e., the coding error of subband coding, can be determined from the sum of the individual subband signals:

$$\sigma_{q,\mathrm{SBC}}^{2} = \sum_{k=1}^{m} c \times 2^{-2} \times R_k \times \sigma_{xk}^{2}$$

If $\sigma_{xk}^{2} = 1/M\sigma_{x}^{2}$ is the same for all subbands, there would be no gain due to band splitting: all subbands are coded with same data rate R_k (bits/sample) = R_{PCM} (bits/pixel), and $\sigma_{q,\mathrm{PCM}}^{2} = \sigma_{q,\mathrm{SBC}}^{2}$ is valid.

Because of the source model, we expect that the output is unevenly distributed in the spectrum; in particular the high-frequency subbands contain only a small output. How the result turns out to be a gain is best explained by the tendency to look at an extreme case: if the total output is contained exclusively in the low-pass channel, a data rate is only required for that channel. As a result of the subsampling in the low-pass path we can assign each sample the M-fold data rate compared to PCM, so that coding error output falls in comparison with the PCM. The gain is obvious.

This simple case is unfortunately not applicable to picture signals because the output is distributed over several subbands. The question is, therefore, how can the bit rate be distributed over the individual channels in order to keep the combined error output as small as possible? This important process of subband coding is called *bit allocation*. The total bit rate per sample R_{SBC} is given by the sum of the bits per sample of the individual subbands R_k, whereby a weighting of $1/M$ needs to be included for subsampling.

$$R_{\mathrm{SBC}} = \frac{1}{M} \sum_{k=1}^{M} R_k = \text{constant} \qquad R_k \geq 0$$

Under this condition, the combination of $\{R_k\}$ must be searched for, which yields a minimal total error $\sigma_{q,\mathrm{SBC}}^{2}$. By that the optimum bit rate for a subband becomes

$$R_k = R_{\mathrm{SBC}} + \frac{1}{2}\log_2 \frac{\sigma_{xk}^{2}}{\left[\prod_{l=1}^{M} \sigma_{xl}^{2}\right]^{1/M}}$$

In the event that $R_k \geq 0$, the band will not be transmitted. Interestingly, the error output is the same in all subbands with this bit allocation. The more output a band shows in relation to the geometric mean of all subbands, the larger its share of the total bit rate. The same relationship is also valid for the bit allocation on the coefficients in transform coding. The gain of subband coding G_{SBC} is defined as the relationship of the coding error output of PCM and subband coding. If we assume an optimum bit allocation, G_{SBC} will be at a maximum and can be represented as the ratio of the arithmetic and the geometric mean of the subband outputs,

$$G_{\text{SBC}|\max} = \frac{\sigma_{q,\text{PCM}}^2}{\sigma_{q,\text{SBC}}^2} = \frac{\frac{1}{M} \sum_{k=1}^{M} \sigma_{xl}^2}{\left[\prod_{l=1}^{M} \sigma_{xl}^2 \right]^{1/M}}$$

We can draw some important conclusions from this formula, which is, unfortunately, only applicable for higher bit rates:

- The more unevenly distributed the signal output of the source is in the spectrum, the larger the gain of subband coding will be.[8] As shown before, nothing is gained if all σ_{xk}^2 are identical.
- With increasing *M,* i.e., with ever finer band splitting, the gain moves toward an upper limit. In Ref. 6 it was shown that the theoretical maximum gain then exactly corresponds with that of DPCM. Exactly the same gain is obtained with transform coding with very large transformation lengths.

It is no wonder that, in theory, the same gain is obtained with all these methods: ultimately all three try to exploit to perfection the linear dependency of the signal. To do this, the following conditions must be fulfilled in the ideal case:

- *Subband coding.* Ideal subband filters and a large number of subbands
- *Transform coding.* Very large block lengths and an optimum transformation
- *DPCM.* Ideal predictor

Under these conditions, the gain of all three methods is the same. Furthermore, the source model used no longer contains any redundancies. The differences are due, in practice, to the usual small bit rates and the limited realization effort; in one of the following paragraphs, transform coding and subband coding are compared with each other, but for now subband coding must show its superiority over DPCM.

7.2.3 Subband coding versus DPCM

In practice, subband coding gives better results than DPCM. One reason for this is the quantizing error with DPCM is fed back in order to prevent the predictors from diverging. That this error becomes larger as bit rates become smaller means that the efficiency of DPCM has a tendency to decrease at small data rates (different considerations toward the efficiency of DPCM).[27,28] Also, the typical coding errors of DPCM—granular noise and particularly those errors resulting from slope overload in picture objects—can be subjectively very disturbing. With subband coding, on the other hand, the visual system can be better exploited: the errors in high-frequency subbands are not as noticeable, so they can be quantized more coarsely.[4] The output in some of the high-frequency subbands is so small that some subbands can completely disappear as a result of quantization. At low bit rates, a lack of sharpness or high-frequency noise occurs, which, in comparison, is found to be agreeable.

7.2.4 Applications for band splitting

What do spectra look like in which band splitting is worthwhile, and into how many subbands must the signal actually be split in order to avoid giving away too much of the theoretical gain? Figure 7.3*a* shows a possible spectrum that is completely flat. In this case there are no more correlations; there is no coding gain from band splitting. In Fig. 7.3*b,* each part of the spectrum is flat after the band has been split in half, and still finer band splitting is superfluous. As long as the spectrum within a subband does not yet approach being flat, band splitting must continue. Figure 7.3*c* shows an unevenly distributed spectrum. Band splitting would be worthwhile, but a large number of subbands are necessary in order to avoid giving away too much of the possible gain. With a spectrum such as that shown in Fig. 7.3*d,* we must ask ourselves whether the effort in band splitting is worthwhile.

Figure 7.3*e* could be the power density of the lines of a picture signal. We recognize that with higher frequencies, the spectrum is already relatively flat. A coarse division is adequate enough here. With low frequencies on the other hand, the gradient falls sharply. A much finer division is recommended in this situation. With picture coding, therefore, we do not frequently perform equidistant band splitting in M subbands of identical size but reduce the bandwidth in lower-frequency ranges. One possibility is for an advancing halving of the respective low-frequency subband, in which it is subjected to a repeated band splitting.

This method is compatible with common filter structures. The band splitting is terminated after a few steps. Because correlations that are still useful remain available in the low-frequency subbands, resulting from the sharply falling output density spectrum,

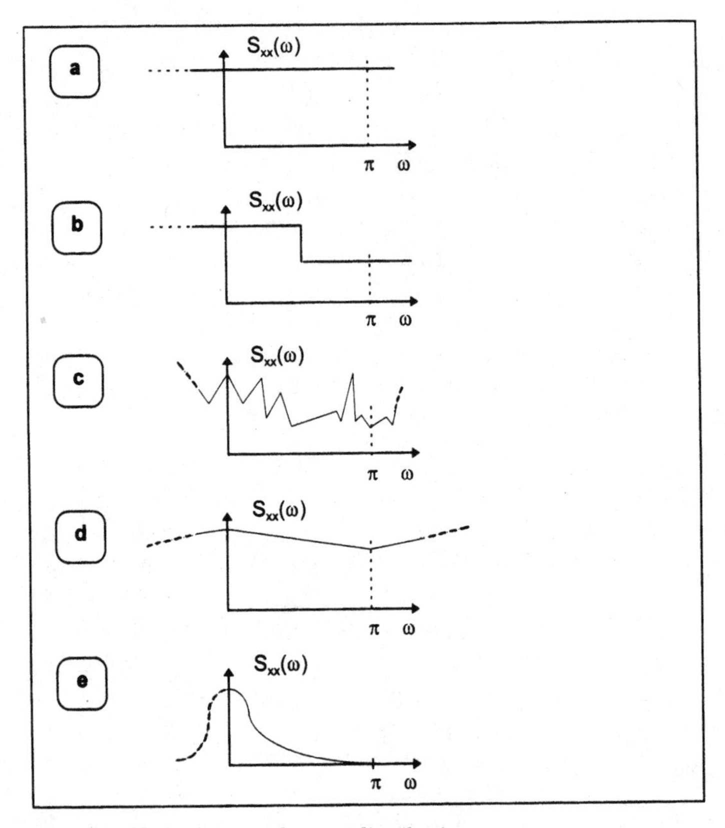

Figure 7.3 Various spectral power distributions.

this is finally transmitted with DPCM[6] or transform coding (e.g., DCT).

7.2.5 A simple example

Figure 7.4 shows a particularly simple system for subband coding. As we (in practice) do not have an infinitely steep filter available, we are obliged to use a certain overlapping of the filter frequency response curves. In this example, the complexity of the filter is minimal: a two-stage transversal filter is used for the low-pass filter as well as for the bandpass filter. The frequency response curve is cosinusoidal in shape in the low-pass branch and sinusoidal in the high-pass branch. The bandwidth of the subsignal is larger than $\pi/M = \pi/2$.

It should be noted that the example given in Fig. 7.4 is identical to a transform:

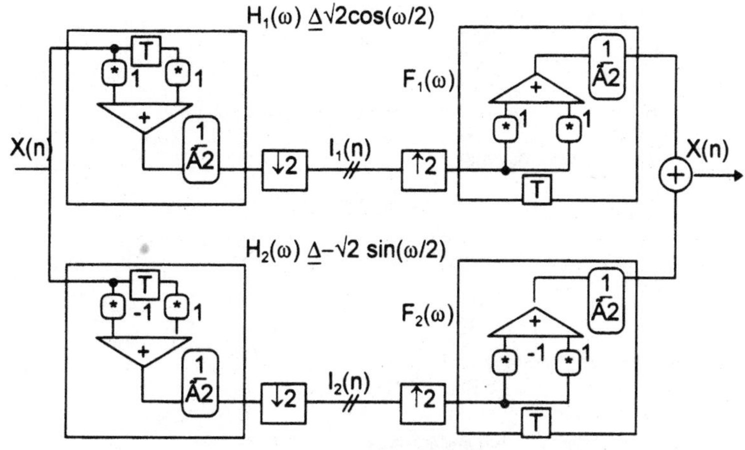

Figure 7.4 A simple example of subband splitting.

$$\rightarrow \begin{bmatrix} I_1(n) \\ I_2(n) \end{bmatrix} = \frac{1}{\sqrt{2}} \times \begin{bmatrix} 1 & 1 \\ 1 & -1 \end{bmatrix} \times \begin{bmatrix} x(2n) \\ x(2n-1) \end{bmatrix} \times \begin{bmatrix} \hat{x}(2n) \\ \hat{x}(2n-1) \end{bmatrix} = \frac{1}{\sqrt{2}} \times$$

$$\begin{bmatrix} 1 & 1 \\ 1 & -1 \end{bmatrix} \times \begin{bmatrix} I_1(n) \\ I_2(n) \end{bmatrix} \rightarrow$$

The effect of the undersampling will be examined in more detail. Figure 7.5 shows the spectrum of the input signal—before and after low-pass filtering with $H_0 = (\omega)$. When nonideal filters are used, there are no open areas in the frequency spectrum that can accommodate the additional bands resulting from subsampling. As a result, low-frequency bands contain components originating from the high-frequency parts; aliasing, therefore, occurs. The low-pass channel is irreversibly disturbed through the incomplete prefiltering. A reconstruction at the decoder seems to be out of the question.

7.2.6 The transformation is also a filter bank

We can interpret the system under consideration in another way: the low-pass filter always gives us the mean value of two pixels, and the high-pass filter the difference between them. Each second pair of values is discarded by decimation. All that is left over is the mean value and the difference for a successive pair of pixels. But this system has been known for a long time: it is TC with the block size of 2 pixels, as

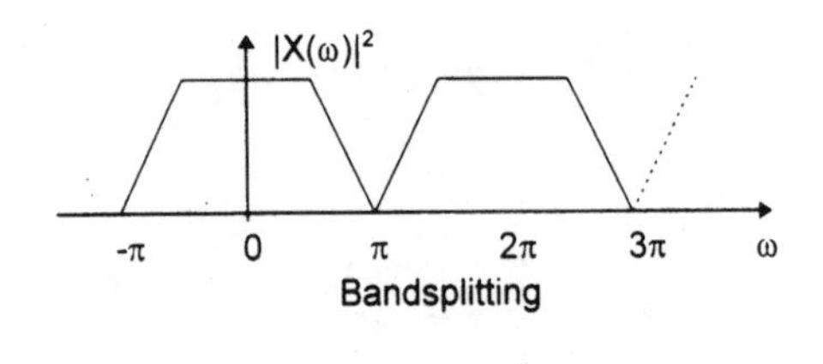

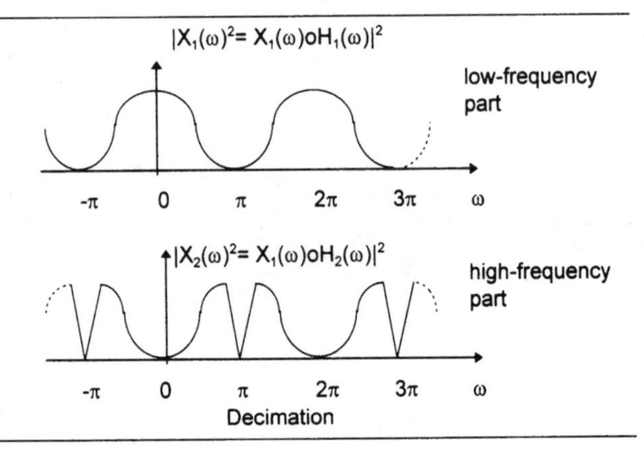

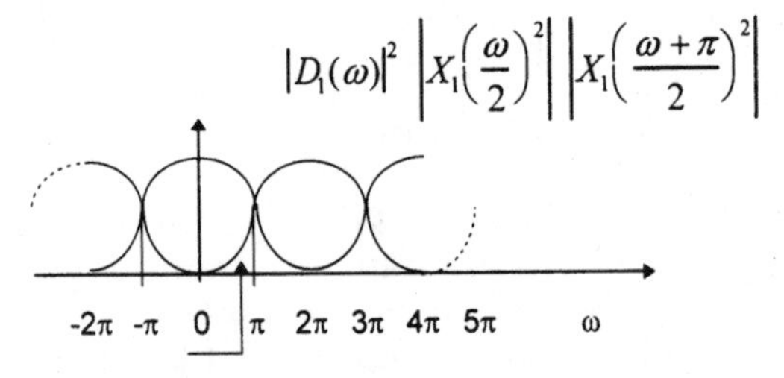

Figure 7.5 Nonideal filters result in aliasing during subsampling.

already described in Chap. 5. Transform coding and subband coding are obviously very closely related. This leads to some interesting conclusions relating to TC:

- Transformation with the block length N corresponds with band splitting into N bands. This band splitting is carried out via N different transversal filters, whereby the coefficients of the k subband filter are found in the k line of the transform matrix.
- Band splitting is, unfortunately, only equidistant.
- The number of filter coefficients is fixed for each subband filter; it corresponds exactly with the transformation length N.
- As the transformation length is, in general, not too long, the out-of-band rejection of the filter is not particularly good. The subbands

contain aliases, and they transport information from other subbands. This results, for the low-frequency coefficients, in a low-resolution picture that is affected by aliasing. If we do without the transmission of the high-frequency coefficients, or quantize them too coarsely, elimination of the aliasing is no longer possible: this results in, for example, blocking with DCT.

- Inadequate separation of the subbands results in incomplete decorrelation. The redundancies of the source are not completely utilized. There is an evident explanation for this condition: simple TC takes no notice of the neighborly relations between blocks that lie close together.
- If we want to improve the quality factor of the filter, filters are required with more stages. That is only possible with TC if we increase the transformation length. However, this gives us an unnecessarily high number of subbands, and the amount of necessary effort increases. An extension of TC which gets around these problems is the *lapped orthogonal transform* (LOT). In this method, a better filter characteristic is obtained by overlapping transformation matrices in the spatial domain.[11]

One characteristic of TC leaves us with a great hope: without quantization TC is a reversible process, although band splitting in alias-containing subbands takes place. The upcoming sections on quadrature mirror filter (QMF) banks and tree structures will show that this works even with general band splitting.

7.2.7 Subband coding versus transform coding

If TC can be interpreted as a filter bank (as with SBC), which advantages, if any, can be obtained for SBC? We have far more freedom when designing the band-splitting diagrams:

- We have the freedom to design steeper filters, by simply increasing the number of filter coefficients. The steeper the filters are, the better the possible gain is exploited (see also Ref. 7). The alias in the subbands is reduced.
- We can adjust the band splitting more flexibly according to the output density spectrum: we are not restricted to equidistant band splitting.
- The number of subbands is independent of the number of filter coefficients. We can thus modify the band-splitting diagram to conform to the concrete signal characteristics better and eventually require fewer subbands.

Without alias-canceling at the decoder, the SBC would be unusable. But this is exactly what was possible with band splitting in the form of TC. Filter banks will now be sought with which (at the neglection of quantization) alias cancellation is possible under less limiting conditions.

7.2.8 QMF banks

We find the solution by deliberately lessening the original demand for ideal filters. We relax the requirement that the filters $H_1(\omega)$ and $H_2(\omega)$ in the coder must be infinitely steep and instead allow them to pass some of the attenuation band. After decimation, a small amount of aliasing occurs. To counter this, we design interpolation filters $F_1(\omega)$ and $F_2(\omega)$, which similarly do not quite eliminate the spectral repetition. Following the addition, the decorrelated signal is composed of a part in the correct spectral position $X(\omega)$ and in a displaced part $X(\omega+\pi)$, which contains the alias:

$$\hat{X} = \frac{1}{2}[H_1(\omega)\,F_1(\omega)+H_2(\omega)\,F_2(\omega)]\,X(\omega) \qquad \text{(desired signal)}$$

$$+\frac{1}{2}[H_1(\omega)\,F_1(\omega+\pi)+F_2(\omega)\,H_2(\omega+\pi)]\,X(\omega+\pi) \qquad \text{(alias)}$$

Perfect reconstruction is possible if $X(\omega) = \hat{X}(\omega)$, not considering a delay. The first step is that aliases from the low-pass channel and the high-pass channel should cancel each other out during addition. The transfer function of the displaced spectrum $X(\omega+\pi)$ must be zero for this condition. There are quite simple conditions for the alias freedom:

$$F_1(\omega) = 2H_2(\omega+\pi)$$

and

$$F_2\,(\psi) = -2H_1(\omega+\pi i)$$

As long as these conditions are met (and they are quite simple), we do not have to worry about the alias any more. In the following, we can ignore the subsampling.

In the next step, it must be decided how, during the design of a low-pass filter for the decoder, the related high-pass filter can be derived. At QMF, we have the following choice:

$$H_2\,(\omega) = H_1\,(\omega+\pi)$$

If $H_1(\omega)$ is a low-pass filter, $H_2(\omega)$ will automatically be a high-pass filter. When choosing the low-pass filter $H_1(\omega)$, we have complete freedom of choice. However, the complete transfer function of the QMF bank $H_{QMF}(\omega)$ must have an amplitude equal to 1 and exhibit a linear phase:

$$H_{QMF}(\omega) = H\frac{2}{1}(\omega) - H\frac{2}{1}(\omega+\pi) = e^{j\omega r}$$

Because of the simple filter design possibilities and guaranteed stability, the finite-impulse-response (FIR) filter (e.g., in the form of transversal filters) is frequently chosen. If the low-frequency subband is to be displayed directly on a monitor, the $H_1(\psi)$ must also be of linear phase, a characteristic which in the video technical world is only reluctantly forsaken. With FIR filters, this can be easily guaranteed through symmetrical filter coefficients, whereas with infinite-impulse-response (IIR) filters, realizing a linear phase is very difficult. (Additional information can be found in Ref. 9.) If we choose a FIR filter, all other filters in the filter bank are FIR filters, and the condition for the phase in the previous formula is automatically met. That only leaves us with one more rule for the amplitude response:

$$\left|H\frac{2}{1}(\omega)\right| + \left|H\frac{2}{1}(\omega+\pi)\right| = 1$$

This formula gives the *quadrature mirror filter* bank its name; the sum of the squares of the low-pass filter and those of the scanning frequency mirrored low-pass filter must be constant. The selection of $H_1(\omega)$ is strictly limited by this means: only the trivial two-stage filter $\cos\omega$ (Fig. 7.4*a*) is allowed for the low-pass filter. We get perfect reconstruction with that, but a cosinusoidal-shaped frequency response curve is not nearly steep enough for band splitting. Alternatively, there are only FIR filters with infinitely numerous stages that meet these conditions. Certainly we have reason to shy away from the effort required for the realization. Realizable QMF banks based on FIR filters and designed according to the stated method have a perfect phase reconstruction, but the amplitude characteristic shows a residual ripple. What should be done now? Mostly, we leave (with good reason) a certain ripple in the total transmission function. Frequently we are satisfied with FIR filters with 16 or 32 stages.

In an emergency, we can compensate for the residual ripple by a correction filter. There are some alternatives available: in Ref. 10, for example, a *conjugate quadrature filter* (CQF) bank has been designed on the basis of FIR filters. The band-splitting systems allow a perfect reconstruction of phase and amplitude, but is the additional design effort worth it? With quantization, it frequently occurs that some of the subbands are not transmitted. The frequency response curve is thereby deformed and the perfect amplitude reconstruction can no longer be maintained. Reference 1 makes mention of an experiment in which it is shown that QMF and QCF give practically identical coding results.

7.2.9 Tree structures for the band splitting of pictures

Instead of a direct splitting into M subbands, QMF banks are often cascaded with two subbands each. Nonequidistant band splitting is, therefore, simple to realize insofar as only the low-pass branch has to be split further (Fig. 7.6). Previously, only one-dimensional band splitting was considered; but pictures are two-dimensional signals. We extend the band splitting in the horizontal and vertical direction—as with DCT—by cascading the splittings in both directions. It is thereby possible to carry out the horizontal splitting completely and then to apply the vertical splitting to the result (Fig. 7.7*a*).

Alternatively, we can immediately follow the halving of a subband in the horizontal direction with a splitting in the vertical direction. A diagram of this process is shown in Fig. 7.7. Such a band-splitting process is used in Fig. 7.8. After an initial band splitting in Fig. 7.8*b*, in both the vertical and the horizontal high-frequency part (right bottom), only a signal (similar to a noise signal) with a small output can be observed. Picture contents, especially those with purely horizontal high-frequency components, can still be seen, which leads us to conclude that the correlation is not complete. Therefore, a further band splitting of these regions is undertaken in Fig. 7.8*c*. A diagram, which in practice has proved itself correct, is shown in Fig. 7.7*c*.[1]

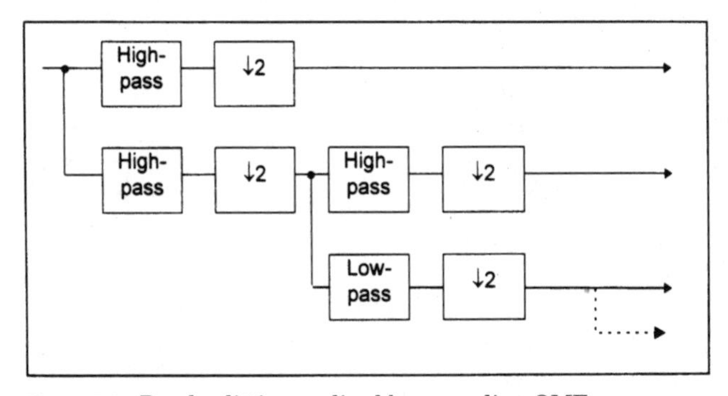

Figure 7.6 Band splitting realized by cascading QMFs.

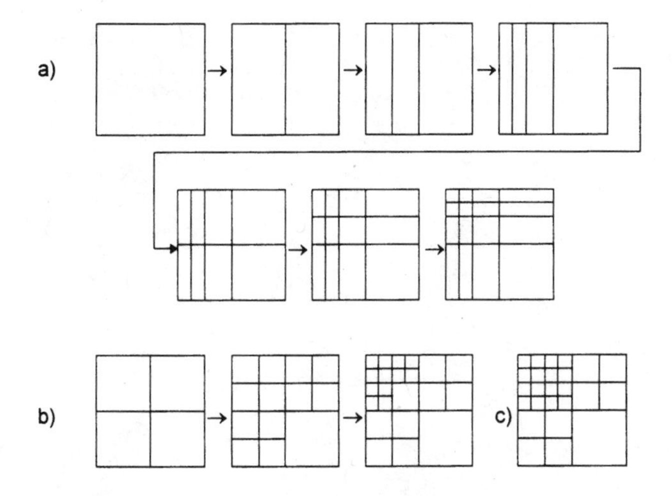

a) separable tree: First complete horizontal separation, then vertical

b) non-separable tree: Alternating horizontal and vertical separation

c) a suitable scheme according to [1]

Figure 7.7 Principal two-dimensional subband splitting schemes.

A small problem remains: depending on the number of filter coefficients, the filters of the QMF bank cause a settling artifact at the edge of the picture, which raises the bit rate unnecessarily and generates reconstruction errors at the picture edges. This can be prevented by continuing the picture to be coded in the reverse order (mirrored) in all directions. The filters can settle in the edge regions so that the middle part can be processed without disturbance. The increased calculation requirement can be compensated by skillful exploitation of symmetries in the QMF coefficients.[1] One method of perfect reconstruction at the picture edges can be found in Ref. 3.

7.2.10 Quantization

As with all other methods, data compression results from the quantization process. In this respect the question arises: How far is it possible to compose a complete picture from the quantized subbands, or does the alias canceling of the QMF banks fail immediately? This is examined in Ref. 1 for those cases where band splitting is not branched out too deeply. It is the case that with an appropriate filter slope (16 taps and above), the error from incomplete alias canceling stays negligible. It is masked by the added quantization noise.

(*a*)

(*b*)

Figure 7.8 Band splitting effects: (*a*) original picture, (*b*) singular subband splitting

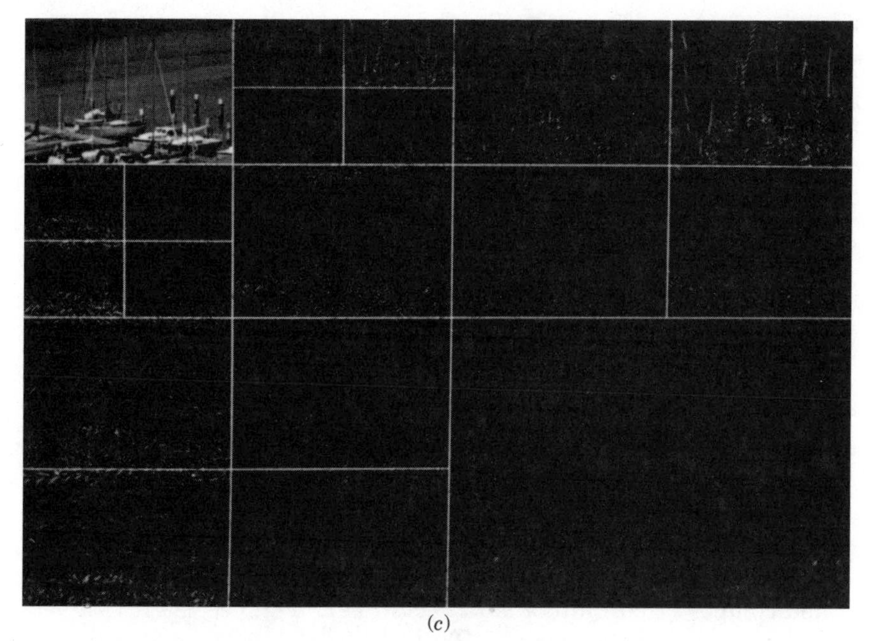

(*c*)

Figure 7.8 (*Cont.*) (*c*) multiple cascaded subband splitting.

Additionally, deterioration of the complete frequency response curve is much more pronounced because of suppression of whole subbands at high frequencies, especially at low bit rates. If a subband has a small output, under certain circumstances, it may not be transmitted at all, even though it may contribute significantly the picture information in a small part of the picture. With adaptive subband coding, this information loss is countered by momentarily donating more bits through a clockwise bit allocation in critical regions of the affected subband.[6]

The statements about quantization errors do not hold true for an increasingly cascaded band splitting. Here it is possible that the resulting filter for the low-pass branch, with increased cascading of filtering and subsampling, shows an impulse response with almost fractal character. The resulting step response from that may lead to noticeable effects on the edges of a low-resolution picture. Luckily, from the relatively young *wavelet theory,* criteria can be derived from which we can conclude whether a filter design shows controllable behavior, despite cascading. Introductory literature on the wavelet theory can be found in Ref. 7.

7.2.11 All dependencies eliminated?

The linear source model used here is not quite true in practice. In picture signals, there are dependencies of a high order, which are not used by the subband coding (and not by DPCM and TC either). We can, however, attempt to exploit these remaining redundancies that are difficult to grasp. There are two possibilities:

- We assume that the scanned values within the subbands show dependencies themselves. The subbands are then treated separately: we cut them into small blocks and carry out a vector quantization on them. With typical picture material the respective code book is given the expected dependencies, in the hope that they will be used during the coding (*intraband VQ*).
- Is it possible that the subbands are not so independent from each other as the source model suggests? Such dependencies can also be caused by nonlinearities during filming. In this case, we can carry out equidistant band splitting, so that each subband contains the same number of samples. We extract one pixel from each subband and compose them into a data vector, before subjecting it to a vector quantization (*interband VQ*).[1]

7.2.12 Application considerations

Subband coding offers great practical advantages because it enables a picture with low resolution to be reconstructed from a mixture of parts of the signal, without a large alias content. This fact can be used in three ways:

- We can provide the low-frequency subbands with better error protection than the higher-frequency subbands. With transmission errors, a recognizable picture will always be generated. It is even possible to reconstruct partly destroyed subbands via nonlinear methods.[2]
- Decoders with limited calculating power or inadequate representation possibilities can at least show low-resolution pictures, in which the higher-frequency subbands are ignored. In order for these pictures to still have an acceptable quality, the overshoot of the low-pass filter as a design criteria must be taken into consideration during construction of the QMF banks. For a hierarchical coding concept with high quality of the high-resolution HDTV picture, the method should be extended in accordance with Fig. 7.9: we transmit the low-pass part as an independent *enhanced-definition TV* (EDTV) picture. Immediately afterward, we calculate the quantization error in the low-pass element. This error picture is

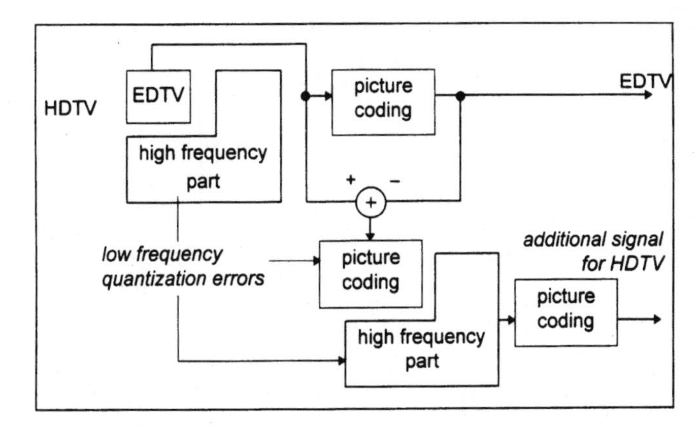

Figure 7.9 System for hierarchical EDTV and HDTV transmission.

substituted for the actual low-pass content into the HDTV picture, and then a separate coding is undertaken of the total. This avoids propagation of the quantization error from the EDTV picture into the HDTV picture. For reconstruction of the HDTV picture, the HDTV receiver uses the EDTV signal and the HDTV residual signal.

- We frequently want to quickly scan the picture data banks and then finally look at a picture in full quality. With this progressive transmission (not to be confused with progressive picture scanning, i.e., noninterlaced), we can quickly and cost-effectively transmit the low-frequency subbands with few bits via, for example, a modem and move through a large catalog. The remaining subbands are only requested when interest is shown in a particular picture.

7.2.13 Subband coding of moving pictures

The previously described intraframe coding does not yet take advantage of the temporal redundancies in a picture scene. It is, therefore, recommended that the application of subband coding to difference pictures be considered.

A simple approach is to carry out a splitting into only four subbands at the coder input. Because there are only a few correlations available from picture to picture, in the high-frequency subbands, a temporal DPCM is not absolutely necessary. These subbands are quantized directly. The low-frequency subband, on the other hand, is subjected to a perfectly normal motion-compensated hybrid DCT.

If we also want to exploit the advantages of subband coding at low bit rates, without reducing the picture quality drastically, we cannot

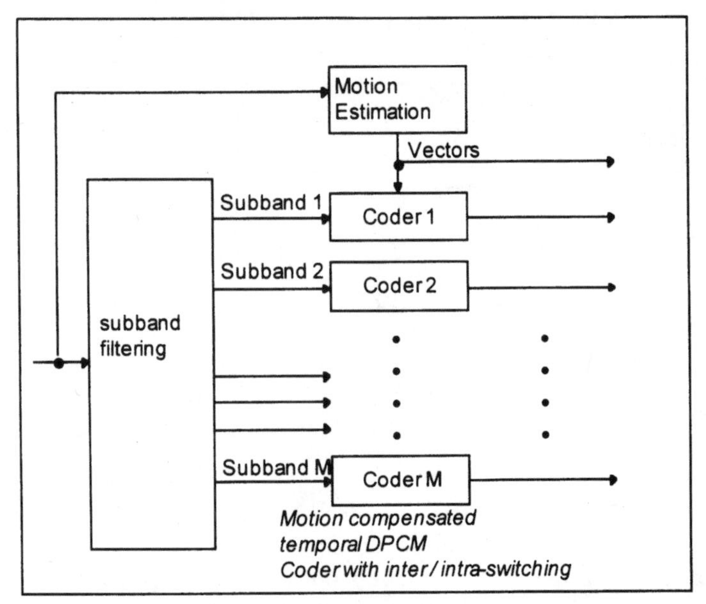

Figure 7.10 Principal diagram for subband coding of differential pictures.

do without the remaining correlations in the high-frequency subbands. These subbands must also undergo a motion-compensated temporal DPCM. With a hierarchical transmission, information from the higher-frequency subbands may not be used for the prediction in the lower-frequency subband. Per definition, there is no HDTV picture known to the EDTV decoder, and therefore no HDTV prediction can be generated. If an HDTV prediction was used in the coder, the result would be a divergence of the predictors. This drift problem can be avoided by using an independent temporal DPCM coder for each subband, (called *in-band prediction*), as illustrated in Fig. 7.10. A second problem is thereby solved: with the temporal DPCM, blockwise switching into the intramode must be performed occasionally. But the subband coding will not work effectively if it is offered a patchwork of difference picture and intrablocks. If the switching between intraframe and interframe in the DPCM coders occurs after band splitting, the band splitting is not affected by it. Obviously, it is better still if, in addition, the sum of all the low-frequency subbands is used for the prediction of the highest resolved picture (called *cumulative coding*).

7.2.14 Subband coding: Outcome and outlook

Subband coding is a method with which efficient coding can be achieved at both high and low data rates. Despite the similarities to

TC, subband coding has some advantages worthy of mentioning: because the lengths of the subband filters are longer with subband coding than the typical block size with TC, they can more efficiently use the redundancy of the picture. On top of this, block effects are avoided. Concepts for hierarchical picture transmission are also supported. However, some cosmetic errors should not be overlooked: because of the long filter lengths of the subbands, the local adaptivity suffers and settling artifacts can lead to distortion at the picture edges. As discussed in previous paragraphs, much more effort is required when combining subband coding with motion-compensated interframe coding, than with TC. For a definitive evaluation of subband coding, we should perhaps wait until more practical experience has been gathered with this relatively young coding technique. Are all the possibilities for data compression with subband coding, TC, or DPCM now exhausted? Unfortunately, all these coding techniques only shine in the light of a very primitive source model.

Pictures are something quite different from low-pass filtered noise. They are composed of edges and surfaces or, more accurately, of objects. The next chapter covers coding concepts that, hopefully, will soon be intelligent enough to recognize objects and, instead of transmitting abstract information signals, be capable of transmitting bit-saving scene descriptions.

7.3 References

1. Westerink, P.H., Subband Coding Images, Dissertation, Univ. Delft, Oct. 1989; s.a.:Westerink, P. H., et al., "Subband Coding of Images Using Vector Quantization," *IEEE Trans. on Communications,* vol. 36, no. 6, June 1988, p. 713.
2. Wang, Y., and V. Ramamoorthy, "Image Reconstruction from Partial Subband Images and its Application in Packet Video Transmission," *Image Communication,* vol. 3, 1991, p. 197.
3. Chaouki, D., et al., "Error-free Image Decomposition/Reconstruction for Subband Coding Schemes," *Image Communication,* vol. 2, 1990, p. 53.
4. Vandendorpe, L., "Optimized Quantization for Image Subband Coding," *Image Communication,* vol. 4, 1991, p. 65.
5. Fischer, T. R., "On the Rate Distortion Efficiency of Subband Coding," *IEEE Transactions on Information Theory,* vol. 38, no. 2, March 1992.
6. Woods, J. W., and S. D. O'Neil, "Subband Coding of Images," *IEEE Transactions on Acoustics, Speech and Signal Processing,* vol. ASSP-34, no. 5, October 1986.
7. Rioul, O., and M. Vetterli, "Wavelets and Signal Processing," *IEEE SP Magazine,* October 1991, p. 14.
8. Jayant, N. S., and P. Noll, *Digital Coding of Waveforms,* Prentice Hall, New York, 1994.
9. Vaidyanathan, P. P., "Quadrature Mirror Filter Banks, M-Band Extensions and Perfect Reconstruction Techniques," *IEEE ASSP Magazine,* June 1987, p. 4.
10. Smith, M. J. T., and T. P. Barnwell, "A Procedure for Designing Exact Reconstruction Filter Banks for Tree-Structured Subband Coders," Paper presented at ICASSP, San Diego, 1984.
11. Malvar, H. S., and D. H. Staelin, "The LOT: Transform Coding without Blocking Effects," *IEEE Transactions on Acoustics, Speech and Signal Processing,* vol. 37, no. 4, April 1989.

Chapter

8

Sound and Audio

8.1 Overview

Sound is the perception of dynamic compression and rarefaction of air pressure through our ears. *Audio* is the electronic reproduction of sound. Generally, humans can hear air pressure variations ranging from 20 Hz to 20 kHz. Loudness, or *sound pressure level* (SPL), is measured in pascals (Pa), with one pascal equal to one newton per square meter (N/m^2). For practical purposes, though, sound is more commonly described as the ratio of one SPL value to another (decibels). As illustrated in Table 8.1, the range of amplitudes that we can hear on a practical basis is about 100 decibels (dB).[1] Decibels are always given as ratios, so there is ample opportunity for confusion if decibels are considered absolute values. Table 8.1 uses dB*A* units, which attempt to give decibel values with the threshold of human audibility set to 0. For audio applications, we are much more concerned about the highest level that the system can handle. So for audio 0, dB is typically about the highest level that the system can handle, and most decibel values in this context are negative.

Our ability to hear low-level sounds is frequency-dependent and is generalized in Fig. 8.1. It should be noted that both axes of the scale

TABLE 8.1 Sound Pressure Levels of Selected Environments

Source	Sound pressure level (dB*A*)
Jet engine, at 10 m from source	150
Rock concert	120
Ambient noise, office	30
Conversation, 1 m from source	60
Quiet breathing	10

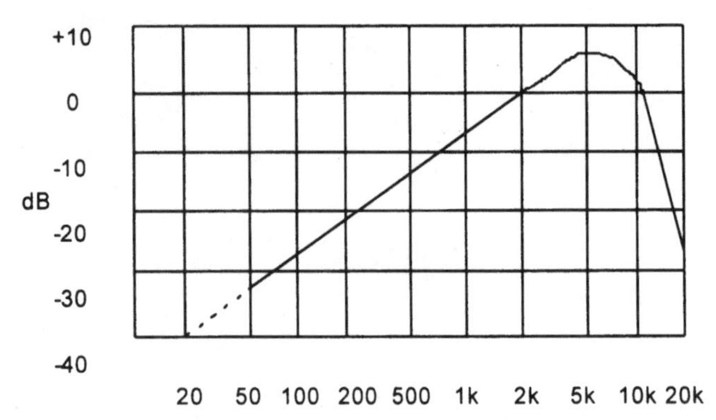

Figure 8.1 CCIR noise-weighting characteristic.

are logarithmic; 6 dB represents a doubling of the audio volume, while a musical octave represents a doubling of audio frequency.

The ear can be modeled as an organ that converts changes in air pressure into signals to the brain with information about power and frequency. The human auditory system uses 26 bandpass filters that overlap one another. The bandwidth of these filters increases as the frequency increases.

Perception is reality. In Chap. 1 we discussed some of the idiosyncrasies of visual perception, such as the Mach effect. In this chapter we will touch on some of the ways that our ears and brain process aural information. Later, we will see how some compression schemes exploit these phenomena.

8.1.1 Auditory masking

It is appropriate to first set out some terminology. The *masker* is the part of the signal that is primarily perceived by the listener, at the expense of the *maskee,* which is comprised of the other elements of the audio signal that are present but not perceived by our brain-auditory system. Because masked elements fall outside of the human realm of perception, they are irrelevant. The masking effect is influenced by four elements:

- Time
- Frequency
- Level
- The nature of the sounds (for example, pure tones versus noise)

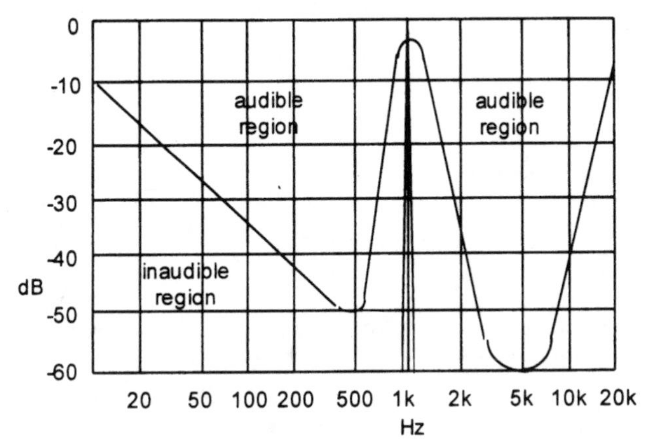

Figure 8.2 Spectral masking properties of a 1-kHz tone.

8.1.2 Spectral masking

Spectral masking occurs when a high-level tone renders lower-level sounds at nearby frequencies inaudible. As Fig. 8.2 illustrates, a 1-kHz tone can mask, for example, a 1.1-kHz tone that is only a few decibels lower. As frequencies get further away from the masking tone, the masking effect diminishes rapidly. As an illustration of this effect, sleigh bells may mask the sound of a high hat but not of a bass drum.

8.1.3 Temporal masking

A sound has an *attack time* (or increase in amplitude with time) and a *decay time* (decrease in amplitude with time). The sound of a violin being plucked has a quick attack and decay, while that of a violin being bowed has a longer attack and decay. In addition, the sound will have a masking effect before its attack and after its decay. Forward masking is in the order of 50 to 200 ms, while backward masking is about one tenth of that duration. Figure 8.3 illustrates the concept.

8.1.4 Distortion

Distortion is a fairly broad term used to describe the extent to which a reproduced signal varies from the original. In politics, we often hear critics complain about the distortion of the facts, and the amount of distortion they perceive is typically proportional to the extent to which their opinion differs from the reproduction device. In audio, as

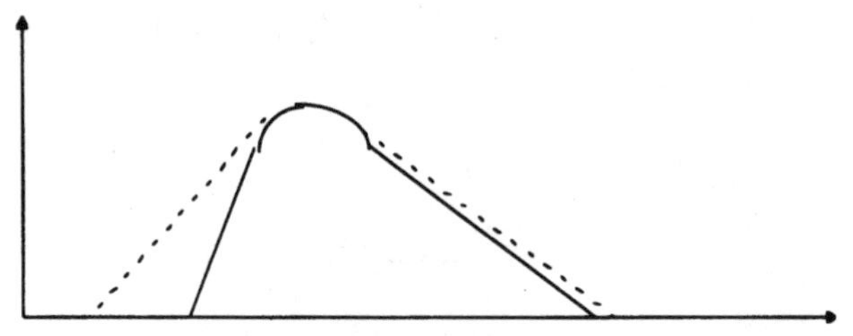

Figure 8.3 The temporal masking effect.

TABLE 8.2 Example of a Five-Point Mean Opinion Score Range

Mean opinion score	Quality scale	Impairment scale
5	Excellent	Imperceptible
4	Good	Perceptible, but not annoying
3	Fair	Slightly annoying
2	Poor	Annoying
1	Unsatisfactory	Very annoying

in politics, distortion is a perceptual phenomenon. We use several mathematical models to try to measure distortion.

8.1.4.1 Subjective measurements of distortion. One subjective assessment is called *mean opinion score* (MOS). Listeners classify the quality of the system by assigning a number from an *N*-point scale. This type of measurement has been used, for example, in selecting an audio compression scheme for HDTV. A common five-point system is given in Table 8.2.

On the one hand, MOS really tries to measure the bottom line in audio reproduction: How does it sound? On the other hand, results vary by listener, test sites, and source material, making it difficult to compare one set of results with another.

8.1.4.2 Objective measurements of distortion. One test that can be calibrated and reproduced is the measurement of the difference between the original and the reproduced signal. Here the drawback is that the absolute amount of distortion may not have much to do with how annoying the distorted sound is.

An example of distortion which we encounter on an almost daily basis, but which is not overly annoying, is clipping. If a pure tone—a sine wave—is put through an amplifier with inadequate dynamic range, it may level off the top and the bottom of the curve, creating a set of odd-numbered harmonics. This type of distortion has a musical relationship between the original (or fundamental) signal and the distortion, so it is not necessarily irritating. Figures 8.4 and 8.5 illus-

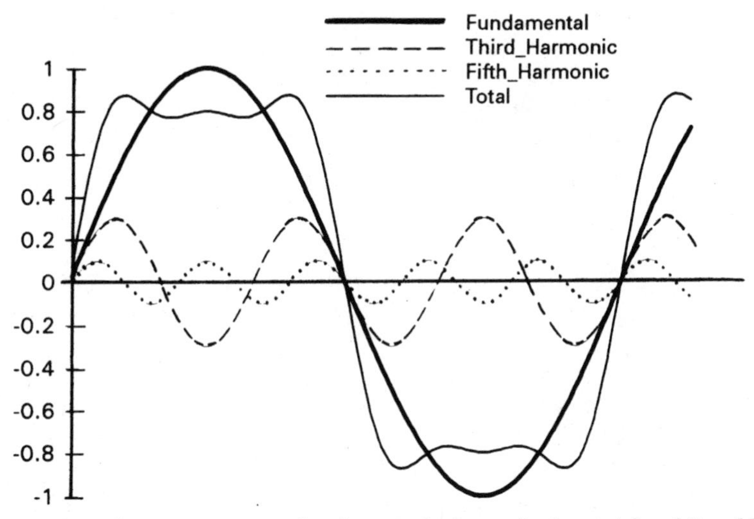

Figure 8.4 A square wave can be characterized as a fundamental and its odd harmonics.

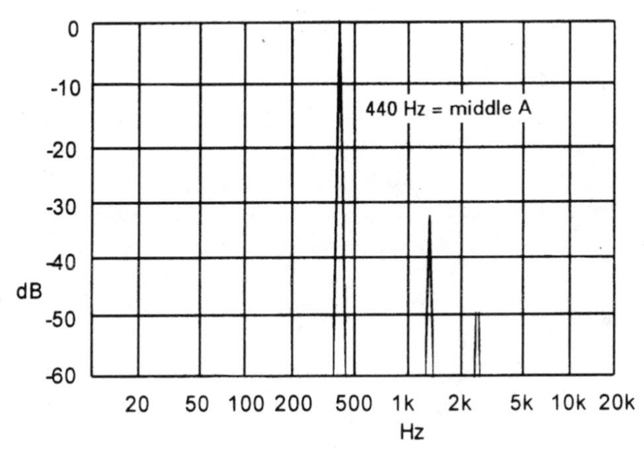

Figure 8.5 Odd-harmonic distortion shown in the frequency domain.

trate this concept. Earlier in this chapter, we discussed other masking effects, which would also weaken any correspondence between measured and perceived distortion.

8.1.5 Channels

While *monophonic* means single source, stereophonic does not mean two sources. *Stereophonic* means giving the three-dimensional effect of auditory perspective. The brain compares three attributes of a sound at each ear in order to locate its source:

- *Amplitude.* If the sound is louder in the left ear than the right, we decide the source is to the right.
- *Phase.* If both ears hear a signal with the same phase, the brain places the sound in the center; if each ear gets a signal that is 180° out of phase with the other, the sound has no directional content.
- *Timing.* Sound travels 1 foot (ft) in 1 ms; if it arrives at the right ear earlier, we decide that the source is to the right.

Because we have two ears, a stereophonic effect can be created with just two speakers in well-controlled environments, such as headphones, or in the case that the listener is in a middle position between two speakers, as shown in Fig. 8.6. If theaters had only one row of seats down the middle, viewers could enjoy a three-dimensional effect with just a left and a right speaker behind the screen. However, theater owners would have a tough time recouping their investment. A more common experience with two speakers is shown in Fig. 8.7. The listener is missing out on full stereophonic effect because she is closer to one speaker than the other.

Localization of sounds can be improved by adding a center channel. The actor's voice will be well-anchored to him as he moves across the screen. Distributing speakers throughout the hall gives the opportunity to add ambient sounds. Figure 8.8 shows the speaker configuration commercialized by Dolby Laboratories in the 1970s. Dolby stereo uses four channels to create the desired three-dimensional effect—

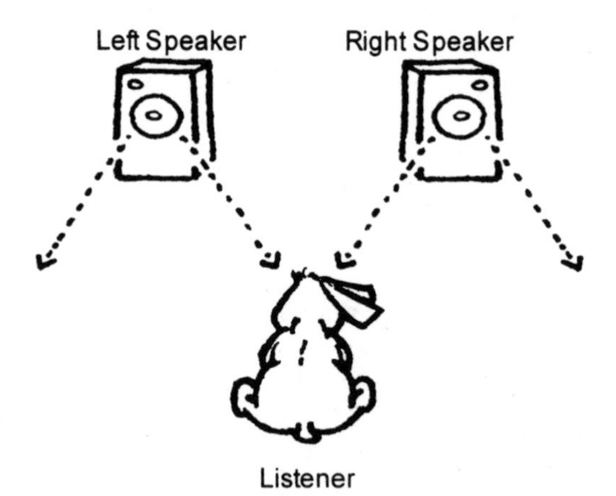

Figure 8.6 A listener enjoying the three-dimensional effect with two speakers.

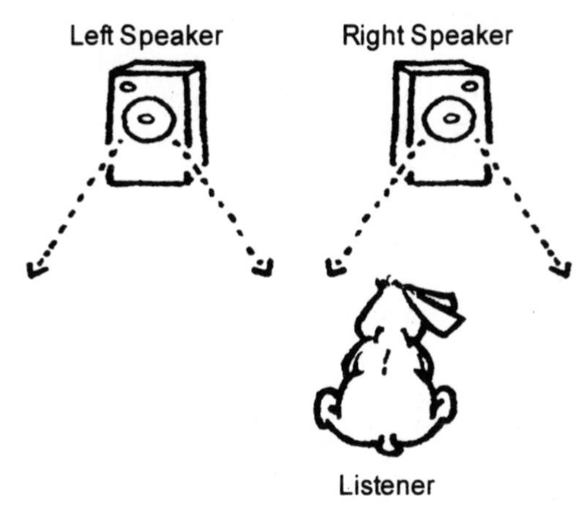

Figure 8.7 A listener missing out on the three-dimensional effect with two speakers.

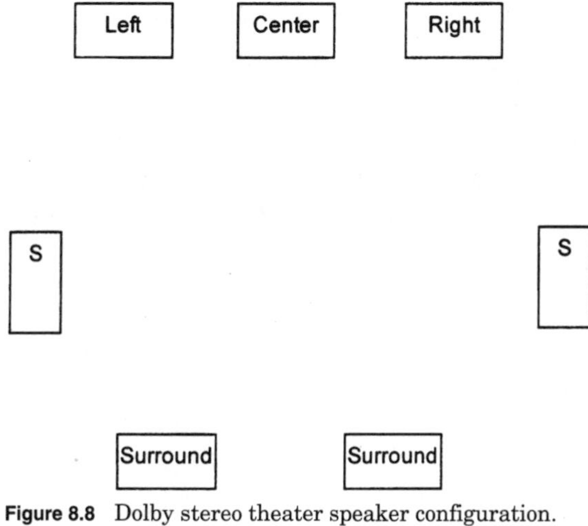

Figure 8.8 Dolby stereo theater speaker configuration.

left, center, right, and surround. Because pairs of audio channels are found in most analog formats (35-mm film, VHS, and stereo broadcast television), the so-called Dolby MP (motion picture) matrix is used to encode the four Dolby stereo channels into two channels.

$$L' = L + 0.7C + 0.7S\times(+90°)$$

$$R' = R + 0.7C + 0.7S\times(-90°)$$

The left and right inputs (L, R) are passed directly through the encoder and form part of the left and right outputs (L', R'), respectively. The center channel is reduced 3 dB and added equally to the left and right outputs (L', R'). The surround channel is phase-shifted $+90°$, reduced 3 dB, and added to the left output channel (L'); it is also phase-shifted $-90°$, reduced 3 dB, and added to the right output channel (R').

One aspect which has led to the widespread adoption of this matrix is its natural backward compatibility with the two-speaker reproduction scheme originally shown in Fig. 8.6. If we return to the lucky listener whose head is in the right space, the center appears to be at the so-called *phantom center.* Because the surround is out of phase it has no directional content; it is perceived as ambient (Fig. 8.9.) In first-generation home decoder systems licensed by Dolby Laboratories (under the name Dolby Surround), the four channels were recovered using the following simple formulas:

$$L = L' \qquad R = R'$$

$$C = L' + R' \qquad S = L' - R'$$

To prevent distracting sounds being perceived through the surround channels, the channel was low-pass–filtered and delayed by about 20 ms.

Declining electronics costs makes increasingly sophisticated systems cost-effective for consumer applications, and surround reproduction is no exception. Dolby now licenses a circuit similar to that used in its theatrical system that employs an active matrix to improve channel separation. The commercial name of this consumer product is Dolby Surround Pro Logic. It will be seen in the following sections

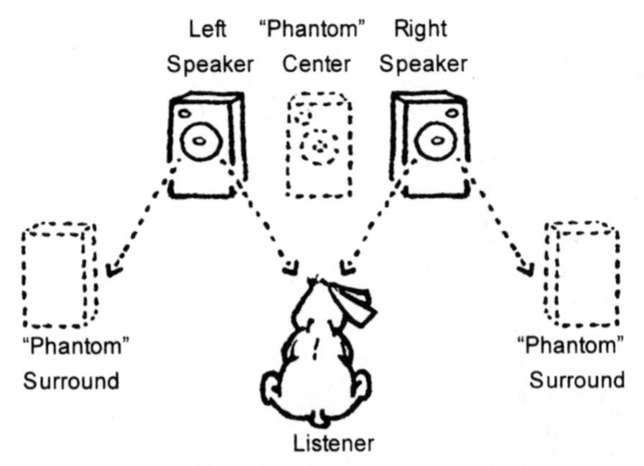

Figure 8.9 A listener enjoying the Dolby stereo effect with two speakers.

that this brief presentation of sound and audio reproduction will help us to understand and evaluate audio compression systems.

8.2 Analog Compression Techniques

While the focus of this chapter is *digital* audio compression, audio compression started with analog approaches, and these same techniques can also be applied in the digital domain. Analog compression techniques seek to improve the perceived dynamic range of reproduced sounds, and earlier in this book we have demonstrated a direct link between dynamic range and the amount of raw data to be handled. Both analog and digital compression methods exploit redundancy and irrelevancy in the input signal as described in Chap. 1.

8.2.1 Preemphasis

The traditional analog methods for audio delivery—FM, phonograph, and magnetic tape—have noise floors that are primarily hissy. When mapped against the low-level sensitivity of our hearing, the result is a "shhhhhhh" sound centered around 5 kHz. Because these schemes were designed in an era of relatively expensive electronics, a very simple scheme is used to reduce the noise perceptibility. All higher frequencies are boosted by applying a preemphasis filter prior to transmission. Linearity is restored by applying a complementary deemphasis filter after reception. Figure 8.10 shows the process graphically.

8.2.2 Wideband companding

With a slight increase in circuit complexity, we can add *wideband companding*. This technique exploits the temporal masking effect. To visualize its implementation, imagine a person with her hand on the playback volume control turning the volume down when there is no sound and turning it up when sound comes along, so that the channel seems noiseless when there is no signal. To try to make the process complementary, we assign another person at the transmitting (or recording) end who turns the volume up when there is no signal and turns it down when a sound comes along. This concept is illustrated in the diagrams of Figs. 8.11 and 8.12.

Unfortunately, there are at least three problems with this approach:

- The first is that the gain reduction in the compressor (and the gain increase in the expander) happens over a finite period of time. This means that the gain will temporarily be too high, resulting in over-

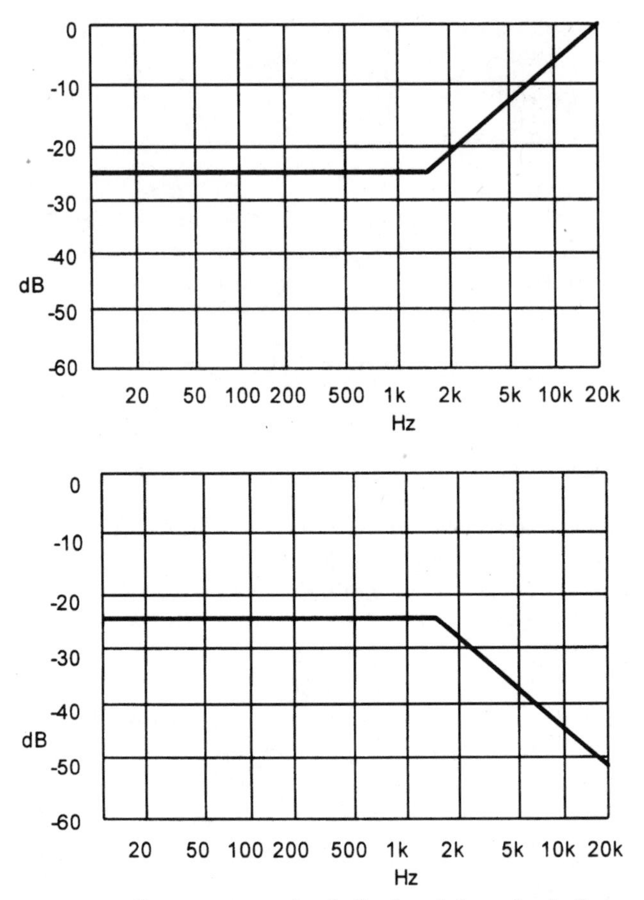

Figure 8.10 Common preemphasis (top) and deemphasis (bottom) curves.

shoots with transient inputs. These overshoots may exceed the level that the channel can handle, resulting in clipping and subsequent mistracking of the expander.

- The second problem is that changing the gain on the sound will distort the signal, possibly producing perceptible problems, such as clicks.
- The third problem is that, as we learned previously in this chapter, a sound masks only nearby frequencies. Thus, if the source sound were a bass drum, we would hear bursts of source-modulated high-frequency noise accompanying each strike of the drumsticks.

In the 1960s, Ray Dolby developed two techniques that solved two of the most annoying problems with companders. He solved the over-

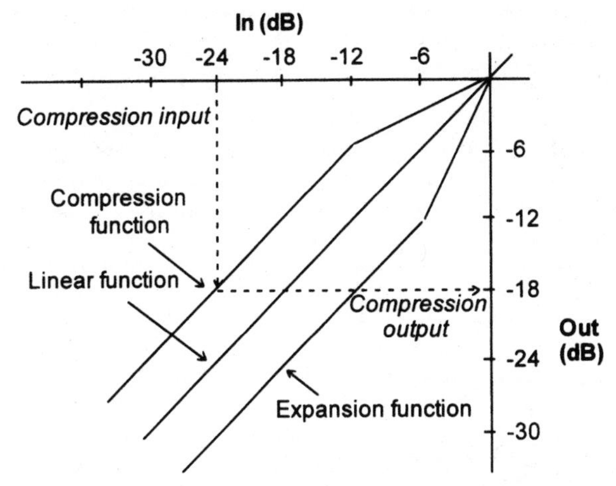

Figure 8.11 A compression system that raises the level of low-level signals by 6 dB.

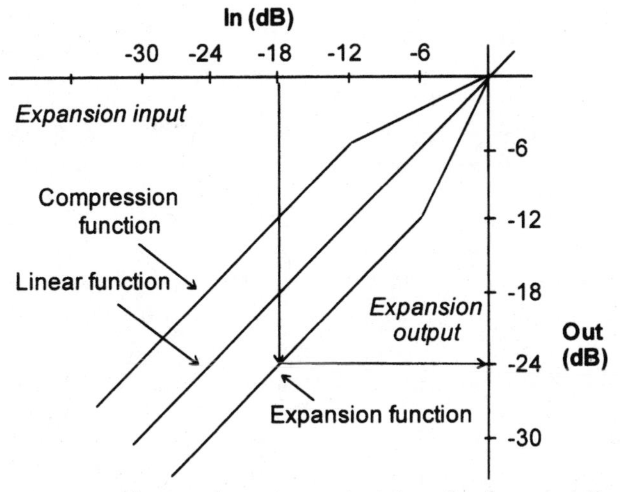

Figure 8.12 The complementary output transfer characteristics of an expander.

shoot problem by creating two signal paths. He solved the modulation of noise by the signal by applying separate companding to different frequency bands.

8.3 Digital Audio Compression

Audio is most typically digitized into PCM. This is how audio is stored on compact disc and digital audio tape (DAT). Because these formats were introduced in a time that digital processing (i.e., digital compression) was more expensive than it is today, PCM was a good choice (for a product introduced in 1983), even though it is inefficient from an information theory point of view. *Adaptive* PCM (APCM) provides an improvement, working directly with the linear PCM signal. *Adaptive differential* PCM (ADPCM) has been discussed earlier in the context of video compression. The bit-rate reduction is achieved by transmitting only the difference elements.

The successful implementation of a digital compression scheme for audio signals involves a number of design considerations, some of which are at odds with one another:

- Overall system complexity
- Available bandwidth for transmission and/or storage
- Audio separation requirements of the application
- Robustness of the coder-decoder system, including the interconnecting link(s)
- Overall cost to produce the system

A number of successful techniques have been introduced commercially, including, for example, *adaptive transform acoustic coding* (ATRAC). This scheme was developed by Sony for use in its magneto-optical MiniDisc (MD). The ATRAC technique employs a combination of subband coding and modified discrete cosine transforms.

An ATRAC encoder accepts a 16-bit PCM input and splits the signal into three bands—0 to 5.5 kHz, 5.5 to 11.0 kHz, and 11.0 to 22.0 kHz. After signal splitting, an appropriate dynamically windowed *modified discrete cosine transform* (MDCT) is applied. The MDCT output frequencies are then allocated into 52 frequency bins. The bandwidth of each bin corresponds to the critical bands for human hearing. This method compresses the data to one fifth of the original input data rate. Figure 8.13 shows a simplified block diagram of an ATRAC encoder.

8.4 Digital Compression Techniques

As of this writing,* the main compression formats in use are based on

- Predictive or ADPCM time-domain coding

*The following material is based on Fred Wylie, "Audio Compression Techniques," in J. C. Whitaker (ed.) *The Electronics Handbook,* Chap. 83 CRC Press, Boca Raton, Florida, 1996, pp. 1265–1272. Used with permission.

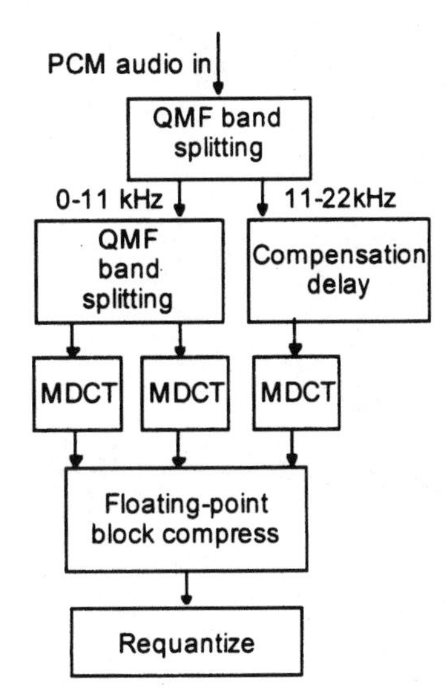

Figure 8.13 Block diagram of an ATRAC encoder.

- Transform or APCM frequency-domain coding

It is in their approach to dealing with the redundancy and irrelevancy of the PCM signal that these techniques differ. Subband coding variations of both these techniques figure highly within the audio industry. The present range of popular algorithms process the PCM signal in narrow, discrete frequency bands, splitting the frequency range of the signal into 2, 4, 32, 256, or 576 subbands, depending on the system being used.

The bit allocation is, then, either dynamically adapted or fixed for each subband. In doing so the frequency-domain redundancies within the audio signals can be exploited, allowing for a reduction in the coded bit rate, compared to PCM, for a given signal fidelity. Because the signal energies in each subband are unequal at any instant in time, this will also lead to spectral redundancies. By altering the short-term coding resolution in each band, according to the energy of the subband signal, the quantization noise can be reduced across all bands. This process compares favorably with the noise characteristics of a PCM coder performing at the same overall bit rate.

On its own, subband coding, incorporating PCM in each band, is capable of providing an improvement in performance compared with that of full-band PCM coding, since both are fed with the same complex, constant-level input signal. This improvement (*gain*) is defined as the ratio of the variations in quantization errors generated in each case while both are operating at the same transmission data rate.

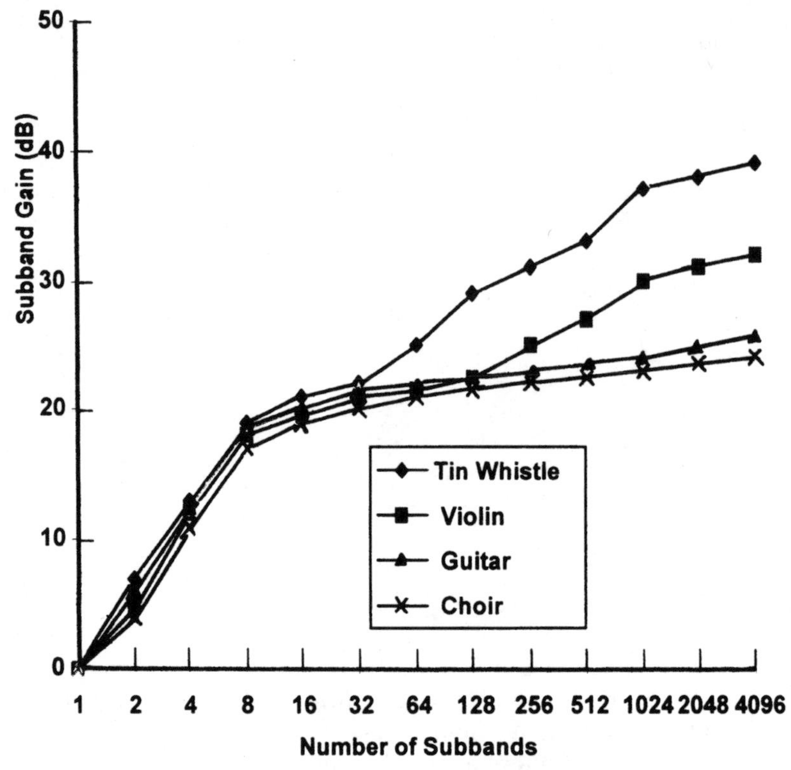

Figure 8.14 Variation of subband gain as a function of the number of subbands for four audio sources.

The gain increases as the number of subbands increases and varies with the extent of the complexity of the input signal.

Figure 8.14 illustrates the variation of subband gain with number of subbands for four essentially stationary, but differing, complex audio signals. In the practical implementations of compression codecs, a number of factors tend to limit the number of subbands employed. First, the level variation of normal audio signals leads to an averaging of the energy across bands and a reduction in the coding gain. The coding or processing delay and computational complexity are two further problems that must also be addressed.

The key issue in the analysis of a subband framework is in determining the likely improvement associated with high-quality audio and in also determining the relationships between that gain, the number of subbands, and the response of the filter bank used to create those subbands.

8.4.1 Subband ADPCM coding

The G722 algorithm was the first commercial quality compression algorithm in this category and was approved by the CCITT in the mid-1970s. Although originally designed to compress 14-bit PCM sampled at 16 kHz, it is still very much in use for 16-bit PCM, 7-kHz bandwidth, speech-only applications. The algorithm processes the PCM signal in two subbands with 6-bit resolution for the 50-Hz to 4-kHz subband, and 2 bits for the other 4- to 7-kHz subband.

Apart from subband coding the other key components of the subband ADPCM algorithm that collectively achieve compression are

- Linear prediction
- Adaptive quantization

The *apt-X100* algorithm (see Fig. 8.15) uses two stages of *quadra-*

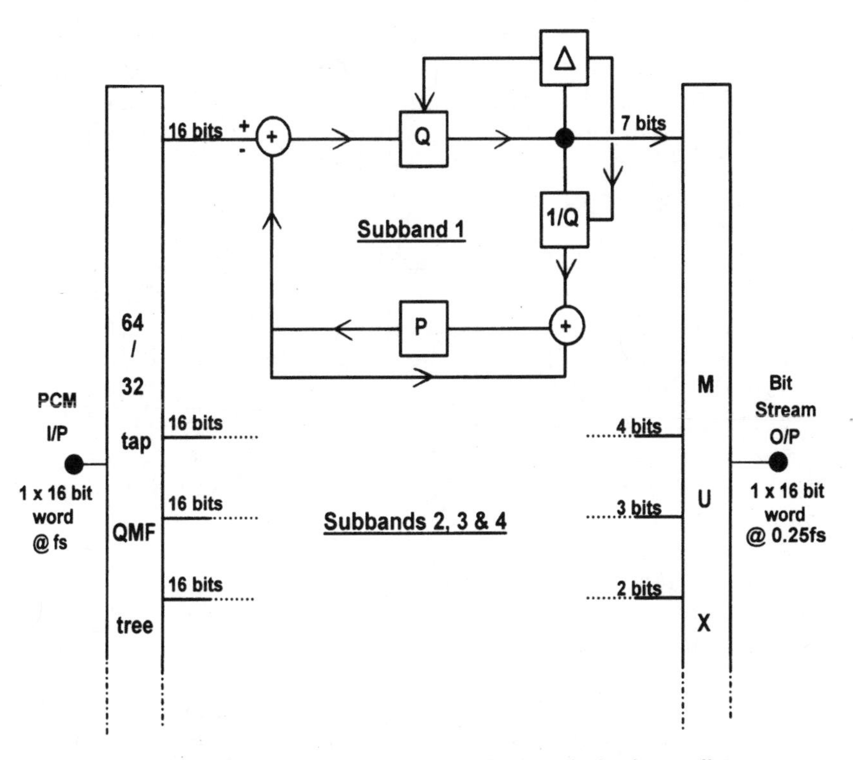

Figure 8.15 Partial block diagram of the apt-X100 coder.

ture mirror filters to initially, in the coder, divide the input linear PCM data into four uniform frequency bands and finally, in the decoder, to reconstruct a linear PCM signal from the compressed data. To reduce interband leakage, which reduces subband gain, the number of filter taps are weighted toward the first of the two quadrature-mirror-filter stages. With this arrangement the complexity and delay time of the filter is manageable, while the subband gain is still quite considerable.

The filter tree operates on four PCM samples as a block. An interrupt routine accumulates the four samples and passes these to the first 64-tap filter stage, where the frequency spectrum of the input signal is divided into low and high subbands. The second 32-tap filter stage, consisting of two similar blocks, then further divides these low and high bands, thus producing four equal and sequential subbands.

8.4.2 Linear prediction

Any reduction in potential gain from the use of just four subbands can be compensated for by linear prediction, which aims to remove spectral redundancies remaining in the subbands. With the inclusion of linear prediction the total performance of the coder can easily match the subband gain associated with the use of many more subbands, without resorting to an increase in filter complexity or coding delay.

Redundancy is removed by subtracting a predicted signal, derived from coder lookup tables, from the incoming subband signal. This action creates a difference signal that is then requantized. If the prediction is accurate enough, then the magnitude of the difference or error signal will be quite small, much less than the original subband signal. This allows the bit allocation to be effectively reduced. The degree to which the subband signal is attenuated is termed the *prediction gain* and represents an improvement in the coded SNR compared with linear PCM.

In the decoder, a signal—identical to that predicted and removed in the coder and similarly derived from lookup tables—is added back again to the now-decoded error or difference signal. This enables the reconstruction of the original linear PCM signal with a minimal loss of information.

The algorithm utilizes backward adaptation of the predictor coefficients, involving no coding delay, allowing the predictor to follow the short-term characteristics of a signal without the necessity to transmit additional side information.

The success of linear predictive coding is highly dependent on the predictability or correlation of the subband signals. Prediction gain is reduced in the higher subbands, and—for most musical instru-

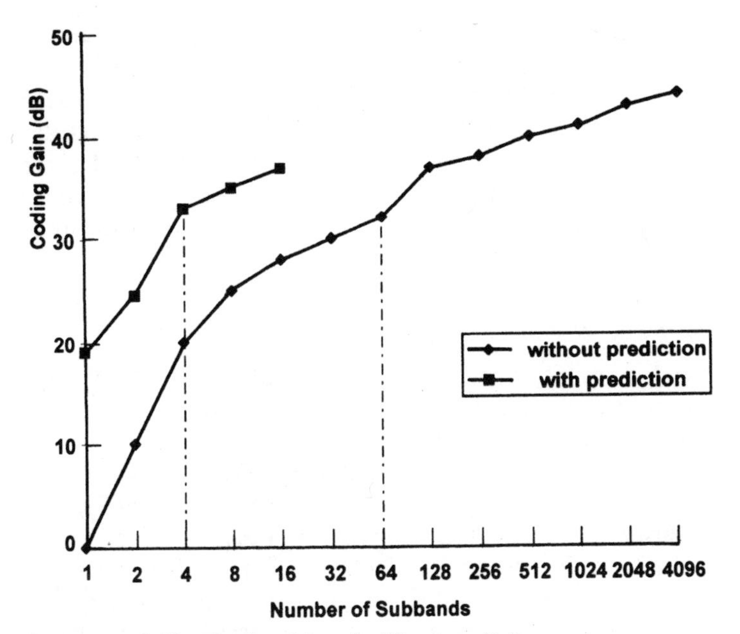

Figure 8.16 Subband gain with and without prediction.

ments—the gain is mainly in the first two bands. For the trombone signal the four-band coder with prediction is seen to give comparable gain to a 64-band coder without prediction (see Fig. 8.16).

8.4.3 Adaptive quantization

Audio exhibits relatively slowly varying energy fluctuations with respect to time, and adaptive quantization exploits this feature by dynamically adjusting the quantizer step size to match the level of the incoming signal. This is a continuous process and provides an almost constant and optimal signal-to-quantization-noise ratio across the operating range of the quantizer.

The apt-X100 coder uses a Laplacian quantizer with backward adaptation that extracts the step-size information from the recent history of the quantizer output. This avoids the problem of estimation delay and range transmission overheads. Because the adaptation is based on reconstructed values and not the original values, the accuracy of the estimating process is limited. An accurate estimation of a complex audio input, for example, containing fast signal transients, is extremely difficult.

Time-domain algorithms implicitly model the hearing process and indirectly exploit a degree of irrelevancy by accepting that the human ear is less sensitive at higher frequencies. This is achieved in the subband derivative by allocating more bits to the lower-frequency bands. This is the only application of psychoacoustics exercised in ADPCM. All the information contained in the PCM signal is processed, audible or not; no attempt is made to remove irrelevant information. It is the unique fixed allocation of bits to each of the subbands 6 and 2 for G722 and 7, 4, 3, and 2 for apt-X100, coupled with the filtering characteristics of each listener's auditory system that achieves the satisfactory audible end result.

8.4.4 Coding delay of apt-X100

The coding delay or processing time is equivalent to 122 PCM samples and is, therefore, dependent on the sampling frequency being used (see Table 8.3). For audio storage and playback and general broadcasting, this time may or may not be critical. However, apt-X100 is ideal for applications such as audio conferencing over ISDN, live outside broadcasting involving reverse program feeds, or off-air transmission monitoring where the functionality of the service is greatly enhanced by the short delay time of this algorithm. It is accepted that a total delay time (processing + propagation) of greater than 20 ms will begin to create difficult echo effects in a two-way audio interchange.

The short-delay performance characteristic of time-domain algorithms compares favorably with some frequency-domain transform algorithms which, in some cases, have processing delays of tens of milliseconds.

TABLE 8.3 Operational Parameters of G722 and apt-X100 Algorithms

Audio bandwidth, kHz	Data rate, kbits/s	Sampling frequency, kHz	A to A delay, ms	Mono	Stereo or dual mono
G.722 Algorithm with 4:1 Compression					
7	64	16	1.4	—	
apt-X100 Algorithm with 4:1 Compression					
7.5	64 & 128	16	7.6	—	—
12	96 & 192	24	5.5	—	—
15	128 & 256	32	3.8	—	—
22.5	192 & 384	48	2.7	—	—

8.4.5 Bit error response

The apt-X100 algorithm is inherently robust to random bit errors, and therefore no protection of the transmitted compressed data is necessary. No audible distortion is apparent for normal program material while speech is still intelligible down to a BER (Bit Error Rate) of 1:10. This resilience to bit errors is the result of a combination of the three coding elements. First, distortions introduced by bit errors are constrained within a subband, and, second, backward adaptive prediction and quantization tend to reduce the significance of the errors by spreading their affect over the trailing window of samples used for the updating.

The impact of a bit error on the subband predictors and quantizers is proportional to the magnitude of the differential signal being decoded at that instant. Thus, if the transmitted differential signal is small, which will be the case for a low-level input signal or for a resonant, highly predictable input signal, then any bit error will have a minimal effect on either the predictor or quantizer in the decoder. The bit error response is therefore well matched to the sensitivity of human hearing, traditionally critical signals being relatively immune to errors.

8.4.6 Implementation of apt-X100

The apt-X100 is a hardware-based fixed algorithm with both the encoding and decoding functions available on a single ROM package. With 2 kilowords of random-access memory (RAM) and 25 kilowords of ROM included in the device, it is possible to store and run two channels (or stereo L & R if applicable) of the encoder and decoder algorithms on this device without the need for external memory. As a result, the hardware design around the apt-X100ED processors is straightforward, requiring normally only the PCM analog-to-digital and digital-to-analog converters and some basic timing logic to complete an audio data compression system. The chip is available in two versions with access times at 33 and 25 ns.

8.4.7 Automatic Synchronization (AutoSync)

In this approach, no prior knowledge of the 16-bit compressed word boundaries are necessary to facilitate the decoding of the compressed data stream. The use of AutoSync enables the compressed data to be handled at both the coder and decoder using only bit timing; no word clock is required, and there is no need for bandwidth-limiting data overheads.

Automatic synchronization is obtained by inserting a unique 10-bit sync word into the compressed audio data stream. This is searched for at the decoder, and once found, establishes the compressed word boundaries for certain multiplexed formats. Insertion occurs once every 128×16-bit compressed words, thus making the length of each frame of data 2048 bits or ($2048/f_s \times 4$) s, including the sync word. Each bit of the sync word is distributed, to a fixed pattern, across 10 preselected words in each frame of compressed data. In each of these words a bit from those representing the second of the lower-frequency subbands is taken to allow for insertion of the sync-word bit. To establish synchronization, even under adverse channel error conditions, three consecutive sync words must be found by the decoder.

8.4.8 Subband APCM coding

The APCM processor acts in a similar fashion to the automatic gain control in the ear, continually adjusting in response to the dynamics, at all frequencies, of the incoming audio signal. Transform coding takes a time block of signal and analyzes it for frequency, energy, and identification of the irrelevancy content. Again, in order to exploit the spectral response of the ear, the frequency spectrum of the signal is divided into a number of subbands and the most important criteria are coded with a bias toward the more sensitive low frequencies. At the same time, using psychoacoustic masking techniques, those frequencies which it is assumed will be masked by the ear are also identified and removed. The data generated, therefore, describe the frequency content and the energy level at those frequencies, with more bits being allocated to the higher-energy frequencies than those with lower energy.

The larger the time block of signal being analyzed, the better the frequency resolution will be, and it will also increase the amount of irrelevancy identified. The penalty is, however, an increase in coding delay and a decrease in temporal resolution. A balance has been struck with advances in perceptual coding techniques and psychoacoustic modeling leading to increased efficiency.

This hybrid arrangement of working with time-domain subbands and simultaneously carrying out a spectral analysis can be achieved by using a dynamic bit allocation process for each subband. This subband APCM approach is found in the popular range of software-based MUSICAM, Dolby AC-2, and ISO/MPEG Layers 1 and 2 algorithms. Layer 3, a more complex method of coding and operating at much lower bit rates, is a combination of the best functions of MUSICAM and ASPEC, another adaptive transform algorithm. (See Table 8.4 for operational parameters.)

TABLE 8.4 Operational Parameters of Subband APCM Algorithms

Coding systems	Compression ratio	Subbands	Bit rate, kbits/s	A to A delay, ms	Audio bandwidth, kHz
Dolby AC-2	6:1	256	256	45	20
ISO Layer 1	4:1	32	384	19	20
ISO Layer 2	Variable	32	192→256	>40	20
ISO Layer 3	12:1	576	128	>80	20
MUSICAM	Variable	32	128→384	>35	20

Additionally, some of these systems exploit the significant redundancy between stereo channels by a technique known as *joint stereo* coding. After the common information between left and right channels of a stereo signal has been identified, it is only coded once, thus reducing the bit rate demands yet again.

Each of the subbands has its own defined masking threshold. The output data from each of the filtered subbands is requantized with just enough bit resolution to maintain adequate headroom between the quantization noise and the masking threshold for each band. In more complex coders, e.g., ISO/MPEG Layer 3, any spare bit capacity is utilized by those subbands with the greater need for increased masking threshold separation.

It is the maintenance of these signal-to-masking-threshold ratios that is crucial if further compression is contemplated in any postproduction or onward transmission process. In extremis, one of the limits for data compression is reached when the quantization noise rises above the masking threshold and is detected as part of the wanted audio and processed as such and is then heard. Out-of-limits harmonic and phase distortion are other limiting factors.

The DCC (Digital Compact Cassette) uses the simplest implementation of subband APCM with the ISO/MPEG Layer 1 algorithm with 32 subbands offering 4:1 compression and producing a bit rate of 384 kbit/s.

The MiniDisc with the proprietary ATRAC algorithm (discussed previously) produces 5:1 compression and a 292 kbits/s bit rate. This algorithm, using a modified discrete cosine transform technique, ensures greater signal analysis by processing time blocks of the signal in nonuniform frequency divisions, with fewer divisions being allocated to the least-sensitive higher frequencies.

ISO/MPEG Layer 2 (MUSICAM by another name) is a software-based algorithm that can be implemented to produce a range of bit rates and compression ratios commencing at 4:1. The Eureka 147 EBU (European Broadcasting Union) project set up in 1987 in Europe to establish the DAB (Digital Audio Broadcasting) formats recom-

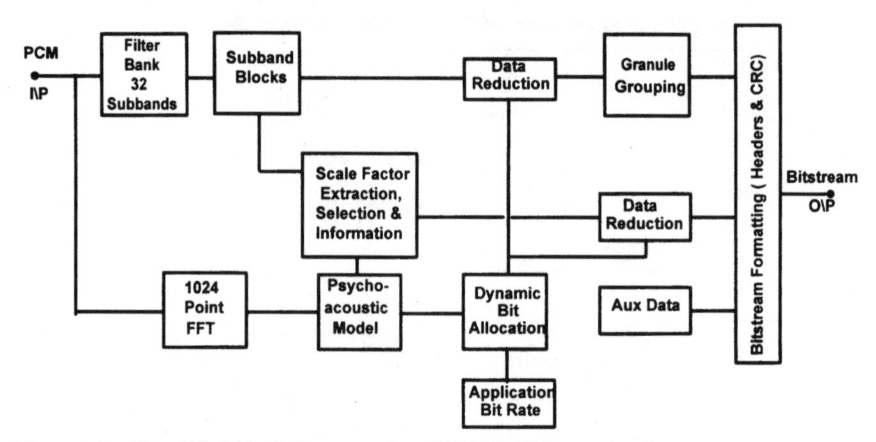

Figure 8.17 Simplified block diagram of an ISO/MPEG Layer 2 coder.

mended using this algorithm for distribution and transmission of 15-kHz stereo at 192 kbits/s and 128 kbits/s, respectively, and lesser rates for speech-only transmissions.

The HDTV Grand Alliance adopted another subband APCM algorithm in Dolby AC-3 for the audio surround system associated with HDTV. Dolby AC-3 delivers five audio channels plus a bass-only effects channel in less bandwidth than that required for one stereo CD channel.

8.4.9 ISO/MPEG layer 2 coding

The difference between MUSICAM and Layer 2 (see Fig. 8.17) is in the structure of each frame of data; Layer 2 has a frame header. This algorithm also differs from Layer 1 by adopting more accurate quantizing procedures and by removing redundancy and irrelevancy on the scale factors.

This subband APCM technique splits the incoming signal into 32 equally spaced subbands using a polyphase analysis filter bank. The bit allocation for each subband is then dynamically controlled by information derived from a psychoacoustic model.

The filter bank, which displays manageable delay and minimal complexity, optimally adapts each block of audio to balance between the effects of temporal masking and inaudible pre-echoes. A 1024-point FFT runs in parallel with the filter bank. The aural sensitivities of the human auditory system are exploited by using this FFT process to detect the differences between the perceptually critical audible sounds, the nonperceptually critical sounds, and the quantization noise already present in the signal; adjustment is then made to the masking threshold, according to a preset perceptual model, to suit the input signal.

The psychoacoustic model is only found in the coder, thus making the decoder less complex and providing the freedom to exploit future improvements in the coder design. The actual quantizer level is determined by the bit allocation, and this is arrived at by realizing the *signal-to-mask ratio* (SMR), defined as the difference between the minimum masking threshold and the maximum signal level. This minimum masking threshold is calculated by the psychoacoustic model and provides a reference noise level of just-noticeable noise for each subband. For Layer 2 this equates to 36 subband samples, each corresponding to 32 PCM samples or, in total, 1152 input PCM samples (Fig. 8.18).

In the decoder, after demultiplexing and dequantizing of the compressed data, a dual synthesis filter bank reconstructs the linear PCM signal in blocks of 32 output samples.

8.4.10 Scale factors

Scale factors are determined for each block of 12 subband samples. When the maximum of the absolute value of these 12 samples has been determined, it is quantized into a word consisting of 6 bits. The data transmission rate for these scale factors is further reduced in Layer 2. Three successive scale factors of each subband of one data frame are examined, and a scale factor pattern is determined. Then, depending on the pattern, up to three scale factors plus an additional scale factor select information data word of 2 bits are transmitted for each subband.

In the case of fairly stationary tonal-type sounds, there is very little change in the scale factors and only the largest one is transmitted. However, with a complex sound the transmission of two or even three

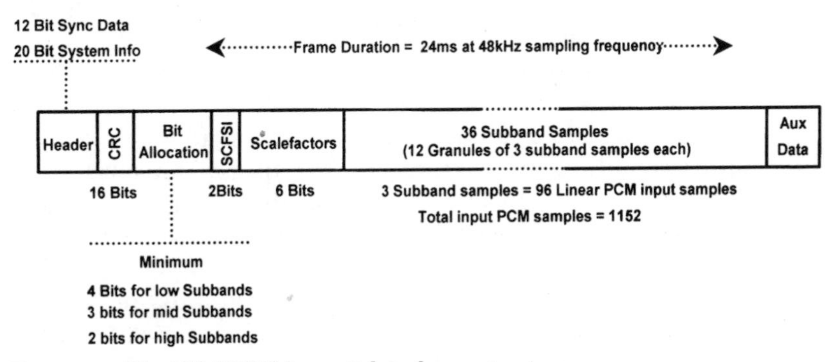

Figure 8.18 The ISO/MPEG Layer 2 data frame structure.

scale factors may be required. Compared with Layer 1, this method of coding the scale factors reduces the bit rate required for them by half.

8.4.11 Audio data bit allocation

After the number of bits required for scale factors, bit allocation information, and other parameters have been determined, the bits remaining are utilized in the coding of the subband samples. This allocation is determined by minimizing the SMR for each subband and the whole frame, and the total is restricted to the number of bits for that frame. Layer 2 has a different allocation of bits across the frequency spectrum, a minimum 4 bits for the lower subbands, 3 bits for the midbands, and 2 bits for the highest subbands. When the audio analysis demands it, this allows for at least 15, 7, and 3 quantization levels, respectively, in each of the three subband groupings. However, each range can cover from 3 to 65,535 levels, and, additionally, if no signal is detected, then there is no quantization. To increase the efficiency of this process, three successive subband samples are grouped into a "granule," and this in turn is defined by only one code word. This is particularly effective in the higher subbands.

8.4.12 Error concealment

In Layer 2 the 16-bit cyclic redundancy code (CRC) word can, if used, provide error detection information for the decoder. This parity-check word, found in each data frame, allows for the detection of up to three single-bit errors or a group of errors up to 16 bits in length. A codec incorporating an error concealment regimen can either mute the signal in the presence of errors or replace the impaired data with a previous, errorfree data frame.

8.4.13 Editing compressed data

Editing compressed audio data can be rather difficult. The linear PCM waveform associated with standard workstations is not available, and the editing resolution of the compressed waveform may or may not be adequate.

The predictive ADPCM approach can allow straightforward editing of the compressed data. For example, with the apt-X100 algorithm an editing window is equal to 4×16-bit samples, which at a 48-kHz sampling frequency creates a time slot of 84 μs, less than a tenth of a millisecond.

The minimum audio sample that can be removed or edited from a APCM-transform–coded signal is determined by the size of the time block of the PCM signal being analyzed. The larger the time block, the more difficult the editing of the compressed data becomes.

8.4.14 Application considerations

No single audio data compression system can optimally cover all the possible applications. Tailoring a system for a specific use is the main aim of most manufacturers, making it crucial that any prospective user is fully conversant with all the pertinent parameters associated with each algorithm. There have been many outstanding successes for digital audio data compression in telecommunications and storage applications. Still, it is important to keep in mind the limitations of the various algorithms and to fully understand their intended applications.

8.5 References

1. Benson, K. B., and J. C. Whitaker, *Television and Audio Handbook for Engineers and Technicians,* McGraw-Hill, New York, 1990.
2. Stallings, W., *ISDN and Broadband ISDN,* 2d ed., Macmillan, New York.
3. Smyth, S., "Digital Audio Data Compression," *Broadcast Engineering,* Intertec Publishing, Overland Park, Kansas, February 1992.
4. Brandenburg, K., and G. Stoll, "ISO-MPEG-1 Audio: A Generic Standard for Coding of High Quality Digital Audio," *Proceedings: 92nd AES Convention,* Audio Engineering Society, New York, 1992; Revised 1994.

Chapter 9

Video Compression Standards

9.1 Setting Standards

As mentioned in the introduction to this book, there are a number of existing and proposed compression systems that employ a combination of processing techniques, much as a chef might employ differing amounts of various ingredients in preparing a meal. Any scheme that realizes widespread adoption can enjoy economies of scale and reduce market confusion. Companies with large market shares, such as Intel or Microsoft, can establish de facto standards that may have an impact greater than that of standards committees.

Timing is an element that is critical to a number of procedures described in this book, including market acceptance of a standard. If a standard is selected well ahead of market demand, more cost-effective or higher-performance approaches may become available before the market takes off. A standard can be similarly academic if it is established after alternative schemes have already become well entrenched in the marketplace. As in the many choices you make in life, caveat emptor.

9.2 Video Compression Systems

In the following sections, we will examine some of the better-known video compression schemes. The examination, while not complete or exhausting, is designed to provide the reader with a starting point for further investigation. The area of video compression is rapidly developing, and numerous technical documents are available from the sponsoring standards-setting organizations.

9.2.1 JPEG

JPEG is enjoying commercial use today in applications where its design goals have proven to be well suited. Because the JPEG standard is the product of a committee, it is not surprising that it includes more than one fixed encoding-decoding scheme. Rather, it can be thought of as containing a family of related compression techniques, from which designers can choose, based on suitability for the application under consideration. The four primary JPEG family members are

- Sequential DCT-based
- Progressive DCT-based
- Sequential lossless
- Hierarchical

Furthermore, additional JPEG schemes have come into practice as a result of adapting JPEG to other environments. We can ascribe these postcommittee variants to capitalism, Darwinism, the third law of thermodynamics, or the lack of muscle that these international committees, such as GATT and the International Standards Organization (ISO), have to make their rules stick. Committee members abide the lack of standardization of their standard by referring to the JPEG specification (as well as the MPEG1 and MPEG2 specifications) as a tool kit of compression techniques. We will first describe the so-called baseline system and then the variants.

9.2.2 Baseline encoder

All JPEG DCT-based coders start by portioning the input image into nonoverlapping blocks of 8×8 picture elements. The 8-bit samples are then level-shifted so that the values range from -128 to $+127$. A fast Fourier transform is then applied to shift the elements into the frequency domain. Huffman coding is mandatory in a baseline system; other arithmetic techniques can be used for entropy coding in other JPEG modes. The JPEG specification is independent of color or gray scale. A color image is typically encoded and decoded in the *YUV* color space with four pixels of Y for each U, V pair.

In the *sequential*-DCT–based mode, processing components are transmitted or stored as they are calculated in one single pass. Figure 9.1 provides a simplified block diagram of the coding system.

The *progressive*-DCT–based mode can be convenient when it takes a perceptibly long time to send and decode the image. With progressive-DCT–based coding, the picture will first appear blocky and then the details will subsequently appear. A viewer can linger on an inter-

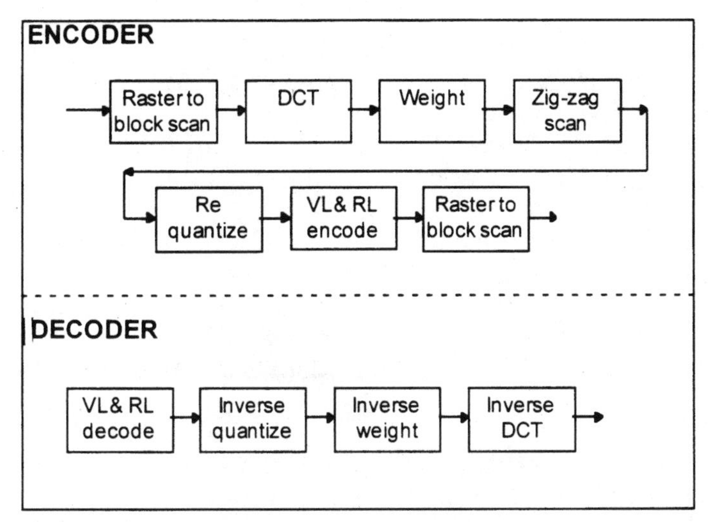

Figure 9.1 Block diagram of a sequential DCT codec.

esting picture and watch the details come into view or may move onto something else, making this scheme well suited, for example, for the Internet.

In the lossless mode, the decoder reproduces an exact copy of the digitized input image. The compression ratio, naturally, varies with picture content. The varying compression ratio is not a problem for sending still photos but presents significant challenges for images that must be viewed in real time.

9.2.2.1 Motion JPEG. Because a picture with high-frequency detail will generate more data than a picture with low detail, the data stream will vary. This is problematic for most real-time systems, which would prefer to see a constant data rate at the expense of varying levels of quality. The symmetry in complexity of decoders and encoders is another consideration in this regard.

9.2.3 MPEG

The Moving Pictures Experts' Group (MPEG) was founded in 1988 with the objective of specifying an audio-visual decompression system, comprised of three basic elements, which ISO calls *parts:*

- Part 1: Systems
- Part 2: Video

- Part 3: Audio

Part 1 describes the system, including audio-video synchronization, multiplexing, and other system-related information. Part 2 contains the coded representation of video data and the decoding process. Part 3 contains the coded representation of audio data and the decoding process.

Just as the great war was renamed when World War II was coined, so the MPEG system, finalized in 1992, was designated MPEG-1 as work began on MPEG-2. The first three stages (system, audio, and video) of the MPEG-2 standards were agreed upon in November 1992. Table 9.1 lists the companies and organizations participating in the MPEG effort.

As of this writing, MPEG was on the verge of broad market acceptance. The techniques of MPEG-1 and MPEG-2 are so similar, and their syntax so extensible, that after noting a few basic differences, we will describe both systems at once.

9.2.4 MPEG-1*

When trying to settle on a specification, it is always helpful to have a target application in mind. The definition of MPEG-1 (also known as ISO/IEC 11172[1,2]) was driven by the desire to encode audio and video onto a compact disc. A compact disc is defined to have a constant bit-rate of 1.5 Mbits/s. With this constrained bandwidth, the target specifications were

- Horizontal resolution of 360 pixels
- Vertical resolution of 240 pixels for NTSC and 288 pixels for PAL and SECAM
- Frame rate of 30 Hz for NTSC, 25 Hz for PAL and SECAM, and 24 Hz for film

A detailed block diagram of an MPEG-1 codec is given in Fig. 9.2. The MPEG standard uses the JPEG standard for intraframe coding by first dividing each frame into 8×8 blocks and then compressing each block independently using the DCT-based method illustrated in Fig. 9.3. Interframe coding is based on MC (Motion Compensation) prediction, but its implementation is far more complex than the one used in H.261 (a standard for the transmission of low-bit-rate visual

*Portions of this section were adapted from G. Lakhani, "Video Compression Techniques and Standards," in J. C. Whitaker (ed.), *The Electronics Handbook,* Chap. 84, CRC Press, Boca Raton, Florida, 1996, pp. 1273–1282. Used with permission.

TABLE 9.1 Participants in MPEG Proceedings

Computer manufacturers
Apple
DEC
Hewlett-Packard
IBM
NEC
Olivetti
Sun
Software suppliers
Microsoft
Fluent Machines
Prism
Audio-visual equipment manufacturers
Dolby
JVC
Matsushita
Philips
Sony
Thomson Consumer Electronics
IC manufacturers
Brooktree
C-Cube
Cypress
Inmos
Intel
IIT
LSI Logic
Motorola
National Semiconductor
Rockwell
SGS-Thomson
Texas Instruments
Zoran
Universities and research
Columbia University
Massachusetts Institute of Technology
DLR
University of Berlin
Fraunhofer Gesellschaft
University of Hannover

communication over telephone lines, e.g., video conferencing). The main difference between the two standards is that the MPEG standard allows bidirectional, temporal prediction. As usual, a block-matching algorithm is used to find the best-matched block that can belong to either the past frame (*forward prediction*) or the future frame (*backward prediction*). In fact, the best-matched block can be the average of two blocks, one from the previous and the other from the next frame of the target frame (*interpolation*). In any case, the

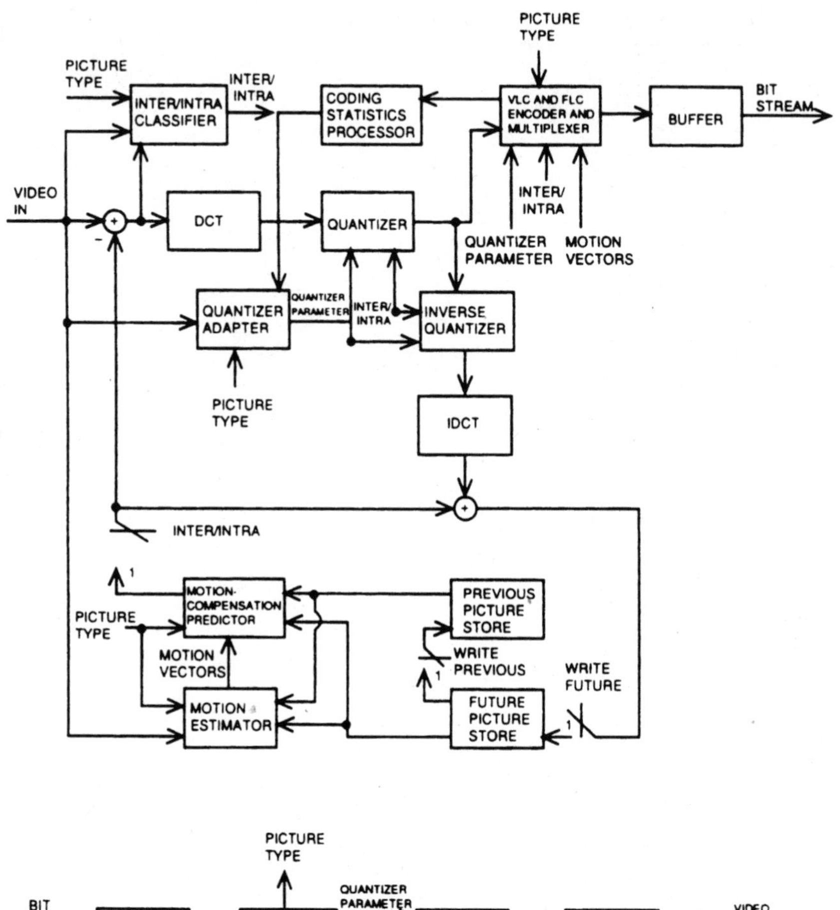

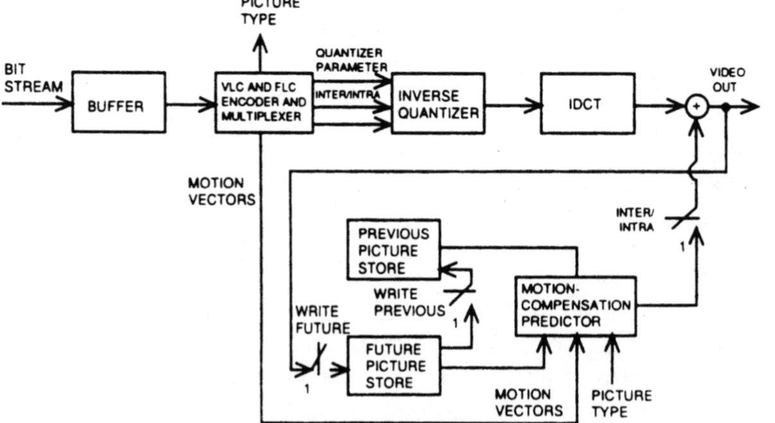

Figure 9.2 A typical MPEG-1 codec. (*Source:* Adapted from Arvind, R., et al., "Images and Video Coding Standards," *AT&T Tech. J.,* 1993, p. 86.)

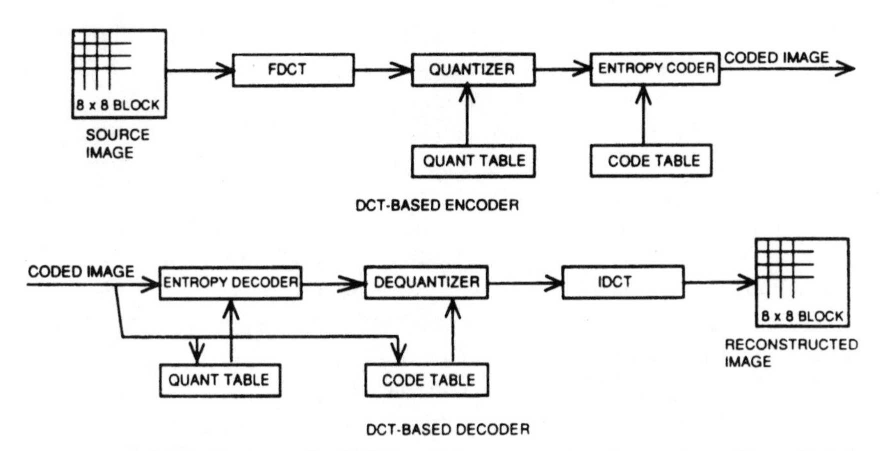

Figure 9.3 A block diagram of a DCT-based image compression system. (*From Ref. 1. Used with permission.*)

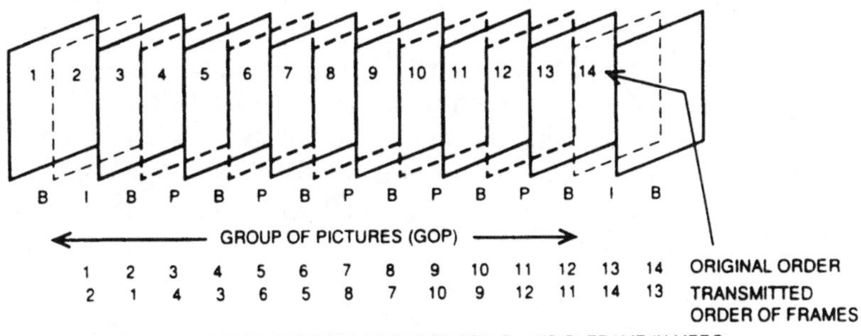

Figure 9.4 Bidirectional prediction in MPEG coding. (*From Ref. 1. Used with permission.*)

placement of the best-matched block(s) is used to determine the motion vector(s); blocks predicted on the basis of interpolation have two motion vectors. Frames that are bidirectionally predicted are never used themselves as reference frames.

Figure 9.4 shows a group of pictures (GOP) of 14 frames with two different orderings. Pictures marked *I* are intraframe-coded. A *P* picture is predicted using the most recently encoded *P* or *I* picture in the sequence. A macroblock in a *P* picture can be coded using either the intraframe or the forward-prediction method. A *B* picture macroblock can be predicted using either or both of the previous or the next *I* and/or *P* pictures. To meet this requirement, the transmission order and display order of frames are different. The two orders are also shown in Fig. 9.4.

The MPEG-coded bit stream is divided into several layers. The outermost layer, called the *video sequence,* contains the basic global information such as the size of frames, bit rate, and frame rate. The GOP layer contains information on fast search and random access of the video data. A GOP can be of arbitrary length. The *picture layer* contains a coded frame. Its header contains the type (*I, P, B*) and the position of the frame in the GOP. The next three layers are the *slice, macroblock,* and the *block layer,* which basically correspond, respectively, to the GOB, macroblock, and block layers of H.261.

The size of the luminance and the chrominance components of the input picture recommended by the MPEG-1 standard are 360×240 and 180×120, respectively, and the recommended frame rate for the coding of the NTSC-compatible TV signal is 29.97 frames per second. The MPEG standard uses blocks of size 16×16 for block matching, and the search is limited to an area of 31×31.

9.2.5 MPEG-2

The primary application of interest when defining the MPEG-2 standard was true TV resolution, which is specified by CCIR-601. This is roughly four times as much picture as MPEG-1 provides. The MPEG-2 standard is a superset, or extension, of MPEG-1. An MPEG-2 decoder should be able to also decode an MPEG-1 stream. It adds to the MPEG-1 toolbox provisions for dealing with interlace, graceful degradation, and hierarchical coding.

While the MPEG-1 and MPEG-2 standards were each specified with specific applications and resolutions in mind, remember that this committee's specifications form a set of techniques that support multiple coding options, including picture types and macroblock types. Picture size can range from 1×1 to 4096×4096, while bit rates can range from 400 bits/s to 100 Mbits/s.

9.2.6 Video source coding

It is normally appropriate to point out at this time that the specifications apply only to decoding, not encoding. The ramifications of this are

- Owners of existing decoding software can benefit from future breakthroughs in encoding processing. Suppliers of encoding equipment have the possibility to differentiate their product by the quality of its encoding.
- Different schemes can be used in different situations. While

TABLE 9.2 Layers of MPEG Video Bit-Stream Syntax

Syntax layer	Functionality
Sequence layer	Context unit
Group-of-pictures layer	Random-access unit: video coding
Picture layer	Primary coding unit
Slice layer	Resynchronization unit
Macroblock layer	Motion compensation unit
Block layer	DCT unit

Monday Night Football will have to be encoded in real time, a film could be encoded in non-real-time, allowing for fine tuning of the parameters via computer.

To allow for a simple yet upgradeable system, MPEG-1 and MPEG-2 define only the functional elements—syntax and semantics—of coded streams. Using the same system of *I, P,* and *B* frames developed for MPEG-1, MPEG-2 employs a six-layer hierarchical structure that breaks the data into simplified units of information. (See Table 9.2.)

The top sequence layer defines the decoder constraints by specifying the context of the video sequence. Its data header contains information on picture format and application-specific details. The second level allows for random access to the decoding process by having a periodic series of pictures; it is fundamentally this GOP layer that provides the bidirectional frame prediction. Intracoded (*I*) frames are the entry point frames, which require no data from other frames in order to be reconstructed. Between the *I* frames lie the predictive (*P*) frames, which are derived from analyzing previous frames and performing motion estimation. These need about one/third as many bits per frame as an *I* frame. Finally, *B* frames, which lie between two *I* frames or *P* frames, are bidirectionally encoded, making use of past and future frames. These frames need only about one ninth the data per frame as the *I* frame.

These different compression ratios for the frames lead to different data rates, such that buffers are required at both the encoder output and the decoder input to ensure that the sustained data rate is constant. One difference between MPEG-1 and MPEG-2 is that MPEG-2 allows for a variety of data buffer sizes to accommodate different picture dimensions and to prevent buffer under flows and overflows.

The data required to decode a single picture are embedded in the picture layer, which is comprised of a number of horizontal *slice layers,* each containing several macroblocks. Each macroblock layer, in turn, is made up of a number of individual blocks. The picture undergoes DCT processing, with the slice layer providing a means of synchronization, holding the precise position of the slice within the image frame.

MPEG-2 places the motion vectors into the coded macroblocks for *P* and *B* frames, and these are used to improve the reconstruction of predicted pictures. MPEG-2 supports both field-based and frame-based prediction (thus accommodating interlaced signals), while MPEG-1 handles frames only. The last layer of MPEG-2's video structure is the *block layer,* which provides the DCT coefficients of either the transformed image information for *I* frames or the residual prediction error of *B* and *P* frames.

9.2.7 Spatial and SNR scalability

Because MPEG-2 was designed in contemplation of handling different picture sizes and resolutions, including *standard definition television* (SDTV) as well as HDTV, provisions were made for a hierarchical split of the picture information into a base layer and two enhancement layers. In this way, SDTV decoders will not be burdened with the cost of decoding an HDTV signal.

An encoder for this scenario could work as follows. The HDTV signal is used as the starting point. It is spatially filtered and subsampled to create a standard resolution image, which is then MPEG-encoded. The higher-definition information could be included in an enhancement layer.

Another use of a hierarchical split would be to provide different picture quality without changing the spatial resolution. An encoder quantizer block could realize both coarse and fine filtering levels. Better error correction could be provided for the more coarse data so that as signal strength weakened, a step-by-step reduction in the picture SNR would occur in a way similar to that experienced in broadcast analog signals today. Therefore, with poorer reception the viewer would experience a graceful degradation in picture quality instead of a sudden dropout.

A *profile* is a subset of the MPEG-2 bit stream syntax with restrictions on the parts of the MPEG algorithm used. A *level* constrains general parameters such as image size, data rate, and decoder buffer size. Most broadcast environments use the main profile with the main level, which gives an image size of 720×576, a data rate of 15 Mbits/s, and a video buffer size of 1835 kbits.

Both MPEG-1 and MPEG-2 specify ISO/IEC 11172-3 audio coding and compression. Table 9.3 attempts to sort out the profiles from the level of enhancements. The current version, MPEG-4, is designed for low-rate audio and video communication applications such as video conferencing and remote sensing. (MPEG-3 was dropped as all its functionalities were met by MPEG-2.)

TABLE 9.3 MPEG Profiles and Levels

		Level			
Profile	Parameter	Low (CIF)	Main (CCIR 601)	High 1440 (HDTV 4:3)	High (HDTV 16:9)
Simple	Image size		720×576		
	Image frequency	—	30		
	Bit rate		15		
Main (4, 2, 0)	Image size	325×288	720×576	1440×1152	1920×1152
	Image frequency	30	30	60	60
	Bit rate	4	15	80	80
SNR scalable (4, 2, 0)	Image size	325×288	720×576		
	Image frequency	30	30		
	Bit rate	3	15		
Enhance Layer 1	Image size	325×288	720×576		
	Image frequency	30	30		
	Bit rate	4	15		
Spatially scalable	Image size			720×576	
	Image frequency	—	—	30	
	Bit rate			15	
Enhance Layer 1	Image size			1440×1152	
	Image frequency	—	—	60	
	Bit rate			40	
Enhance Layer 2	Image size			1440×1152	
	Image frequency	—	—	60	
	Bit rate			60	
High (4:2:2, 4:2:0)	Image size		352×288	720×576	960×576
	Image frequency	—	30	30	30
	Bit rate		4	20	25
Enhance Layer 1	Image size		720×576	1440×1152	1920×1152
	Image frequency	—	30	60	60
	Bit rate		15	60	80
Enhance Layer 2	Image size		720×576	1440×1152	1920×1152
	Image frequency	—	30	60	60
	Bit rate		20	80	100

9.3 Timing and Synchronization Using MPEG-2 Transport Streams*

Part of the power and quality obtained with MPEG-2 video is the result of its use of temporal compression, resulting in forward-predicted pictures and bidirectionally predicted pictures. Timing information is provided in the compressed data at three different levels:

- Compression encoding

*This section is based on D. K. Fibush, "Timing and Synchronization Using MPEG-2 Transport Streams," *SMPTE J.*, SMPTE, New York, July 1996, pp. 395–400. Used with permission.

- Elementary stream packetization
- Transport stream formation

Program clock reference (PCR) time stamps are periodically added to the transport stream packets to provide overall timing synchronization. The PCRs are expected to arrive at the system receiver at regular intervals so the receiver may reconstitute the 27-MHz system clock. This requires a software-driven PLL, which is equivalent to vertical sync genlock, further complicated by the fact that the PCRs need only arrive 10 times per second.

A simplified block diagram of the MPEG-2 television compression system is shown in Fig. 9.5. Television nominally consists of audio and video; however, the system as shown includes data and control signals, and so it may be thought of as a multimedia system. For completeness, analog television inputs to the system are shown, although many facilities are likely to be primarily digital. Video inputs are analog NTSC, PAL, component (e.g., Betacam, M-II, or EBU N-10[3]), or high definition per SMPTE 274M.[4] Video and audio are transformed to standard digital formats by analog-to-digital converters. Digital audio would generally be directly accepted by the audio coder, whereas digital video is more likely to use preprocessing in order to meet the MPEG-2 (Rec. 601[5]) component video input requirement. Composite inputs must be decoded, although that function is often included in an MPEG-2 encoder-compression product.

There are detailed MPEG-2 documents covering video coding, audio coding, and system data formatting.[6] Digital data from each of the coders and other sources is first formed into *packetized elementary stream* (PES) packets and then multiplexed and reformatted into a *transport stream* (TS). The TS uses fixed-length packets and may carry data for a multiplicity of programs. The other similar channels would be for inclusion in the transport stream of other independent programs. A program source is likely to include multiple audio sources and possibly multiple video, data, and control sources. The

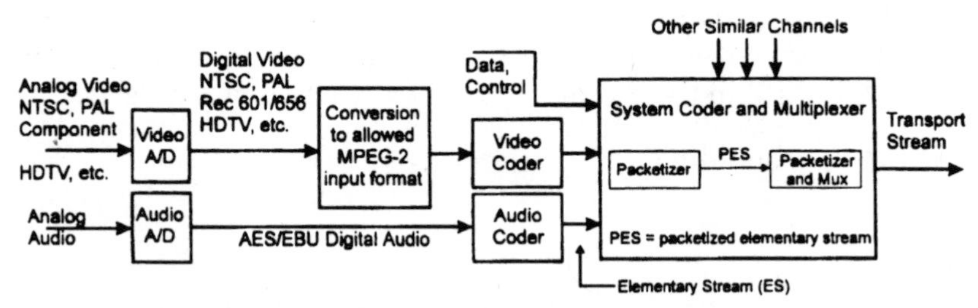

Figure 9.5 Basic MPEG-2 television compression system. (*From Ref. 2. Used with permission.*)

transport stream, with its fixed-length packets of 188 bytes each, is specifically designed for long-range transmission.

9.3.1 MPEG-2 video compression

A complete description of MPEG-2 compression methods may be found in a number of articles.[7,8] The output data resulting from the compression process are known as an *elementary stream* (ES). For the purpose of understanding system timing, only the various types of pictures need be considered:

- *I* pictures with no temporal compression
- *P* pictures using only forward prediction
- *B* pictures that are bidirectionally predicted

A powerful option in MPEG-2 encoding is that the pictures may be either fields or frames, selected by the encoder during the compression process.

The most basic timing information, the temporal reference, is included in the data during the compression process based on the GOP concept shown in Fig. 9.6. A GOP begins with an *I* picture and includes all the pictures up to the next *I* picture with the temporal reference starting at zero and incremented at each subsequent picture. In MPEG-2 there are only eight allowed frame rates, and they must all be locked to the 27-MHz reference. Therefore, elementary streams may be decoded knowing only the specified frame rate and the temporal reference. An arbitrary 27-MHz clock can be used to reconstruct the Rec. 601 video for data stored in some manner, such as on a hard disk. Additional timing information is added to the data

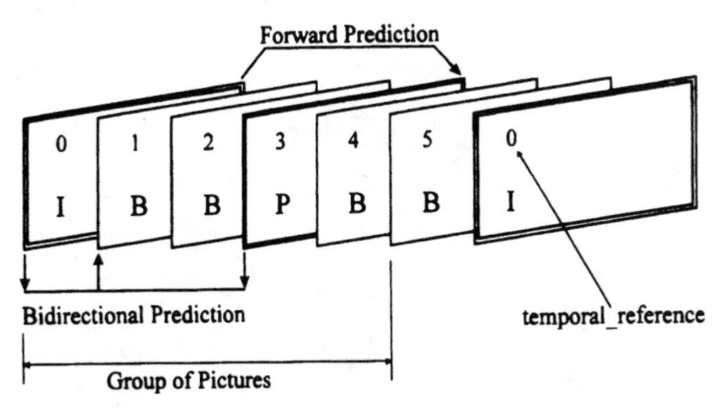

Figure 9.6 The group-of-pictures concept for MPEG-2. (*From Ref. 2. Used with permission.*)

as the transport stream is formed to provide for correct real-time system operation.

As the video is encoded, the pictures will be placed in the data stream in a different order than they were received (only when *B* pictures are used and the GOP is longer than two, which is the usual case). Because of the compression applied to each picture, the amount of data required to transmit that picture is variable, differentiating the timing of the compressed data flow from that of the uncompressed video beyond that of their respective bit rates. Reordering of the pictures (frames) is based on the requirements of temporal compression. Because the *B* pictures are bidirectionally predicted, their compression cannot take place until the next anchor picture (*I* or *P*) has been processed. The net result is a time sequence of data for the compressed pictures (frames in this case), as shown in Fig. 9.7.

9.3.2 MPEG-2 transport stream

Processing signals to produce the transport stream are shown in the block diagram of Fig. 9.5. A conceptual diagram of the packetizing of video data is shown in Fig. 9.8. MPEG-2 compression starts with Rec. 601 video and develops a lower-bit-rate structure of elementary stream data consisting of, among other things, video sequences, groups of pictures, and single pictures. Although there is structure to the compressed video elementary stream, it is generally utilized as a single large block or file of data.

The first process in the development of the transport stream from the elementary stream data is a logical division of the data to form the PES. This is logical because the PES packets need not actually exist in the data-processing equipment. These are large packets, up to 64 kbytes in size or longer, which may contain a complete compressed video picture. Headers for the PES packets contain timing information specific to the one elementary stream. These are, among other

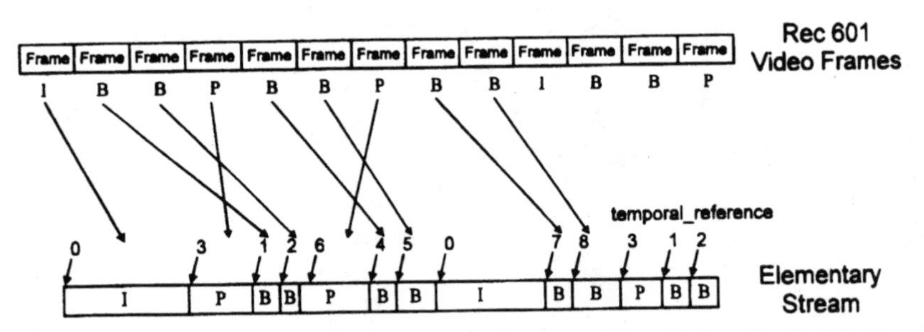

Figure 9.7 Time sequence of compressed video pictures. (*From Ref. 2. Used with permission.*)

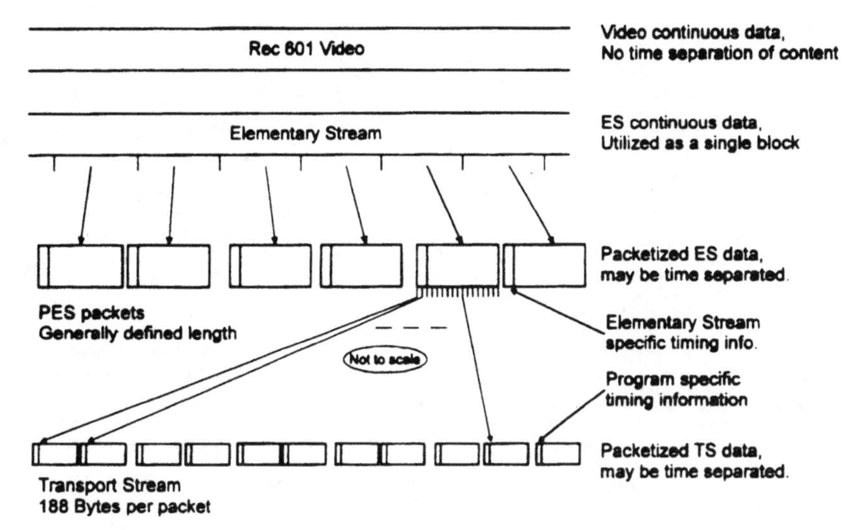

Figure 9.8 Conceptual diagram of video compression and packetizing. (*From Ref. 2. Used with permission.*)

things, the *presentation time stamp* (PTS) and the *decode time stamp* (DTS), which are used to provide directions to the decompression equipment. PES packets are not considered to be a continuous block of data; they may be separated in time. In fact, consideration has been given to using them directly in transmission systems although the preferred method for large-packet transmission is the MPEG-2 program stream.

Transport stream packets are subdivisions of PES packets with additional header information. Two of the most important types of information contained in the transport stream packet header are the *packet identification* (PID) and the PCR. Every transport packet has a PID, which is used to reconstruct the variety of separate data streams that may be extant in one transport stream. The PCR contains timing information that is used to synchronize the 27-MHz clocks in the encoder and decoder. Transport stream packets are exactly 188 bytes, with a minimum header of 4 bytes and a longer header when the PCR is present. Transport stream packets for a single elementary stream may be separated in time specifically for interleaving additional transport stream packets, although there are other cases where they may be time-separated such as for an automatic teller machine transmission.

A program consists of one or more elementary streams with the same timing reference; that is, one set of PCRs provides the timing information for all types of data associated with that program, and any individual timing reference, such as for digital audio and video,

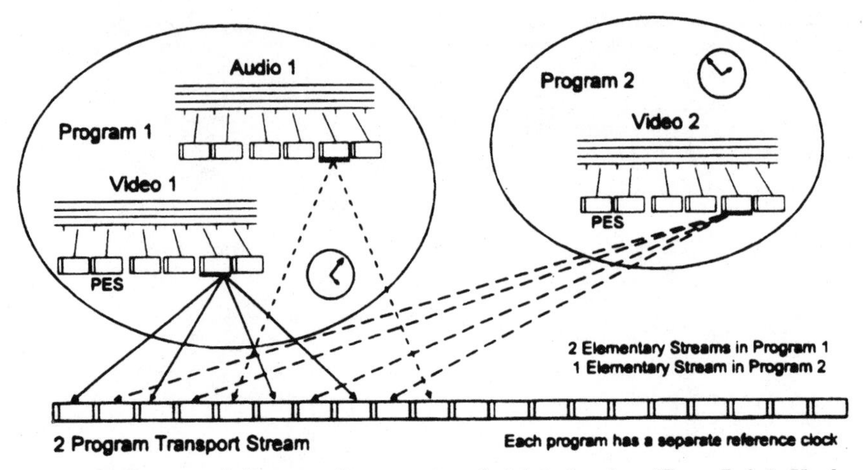

Figure 9.9 Conceptual diagram of transport packet interleaving. (*From Ref. 2. Used with permission.*)

must be derived from the same master clock. The interleaving of transport stream packets from more than one program is shown in Fig. 9.9. Program 1 contains two elementary streams, one video and one audio. Program 2 contains only a video elementary stream. Each program has its own 27-MHz clock, depicted by the clock-on-the-wall icon in the drawing. A complete transport stream has a constant bit rate, although the individual programs or elementary streams may have a variable bit rate. Null packets (PID 8191 = $1FFF_h$) are used to pack the transport stream with enough data to maintain the constant bit rate.

Programs with independent timing information are a very important part of the MPEG-2 system concept. This allows programs to be derived from completely independent sources and facilitates the multiplexing of programs with different, and varying, bit rates into one transport stream. More than one program in a single transport stream is the basis for digital television transmission over radio frequency channels of a fixed bandwidth. Using statistical multiplexing, higher bit rates are provided for transmission of the more difficult program material. When appropriate, the statistical multiplexing may be manually influenced, for example, to give a higher bit rate to a sports event running in the same transport stream as an old movie.

A transport stream may contain a large number of programs, and each program may contain a large number of elementary streams. Therefore, a system is required to keep track of all the different data streams and their related PIDs. This is accomplished with two mapping mechanisms:

- *Program association table (PAT).* Every transport stream must have a PAT, which is always PID 0. This table is required to identify the PID numbers for the table(s) defining each program.
- *Program map table (PMT).* There is a PMT for every program in the transport stream. This table states the PID number for each elementary stream associated with a specific program.

9.3.3 System time stamps

All timing information for one program is derived from a single 27-MHz system reference clock at the source of elementary channels. Because the 27 MHz relates directly to video scanning rates, video is usually the source for the clock. Digitizing of any other channels, including the audio, must use a sample rate derived from the system reference clock for that program. If the source of the video were a free-running clock in a hand-held camera, that clock would be the reference for all operations related to that program, completely through the system to the final display of the picture. Generally, the system reference clock is a stable sync pulse generator to which all the channel sources are locked.

The 27-MHz system reference is informally called the *clock on the wall* because time stamps added to the data stream are based on clock time for a specific function. Figure 9.10 shows how the clock on the wall is used in forming the transport stream. Rec. 601 frames are identified as an *IBBPBBP...* group of pictures by the encoder algorithm and compressed, with temporal-reference added, into a sequence of *IPBBPBB....* Two time stamps are added in the logical formation of the PES packets. Although the time required to transmit the data for each picture will vary, the PTSs will have exactly the correct clock-on-the-wall time for presentation at the final display corresponding to the video program frame rate. DTSs, which are also added to aid in decoder operation, are not associated with *B* pictures because these bidirectional pictures are always decoded as soon as all the data are available.

The PTS and DTS time stamps are essentially helper signals for the decoder and do not have a required frequency of occurrence. The PCR is required for system operation specifically to synchronize the encoder and decoder system reference clocks. PCRs are contained in the transport stream packet headers at least 10 times per second. Some system designs specify a more frequent occurrence, such as 25 or 30 times per second, corresponding to the video frame rate.

The PCR is a 33-bit base specifying cycles of a 90-kHz clock with a 9-bit extension specifying cycles of the 27-MHz clock, providing a

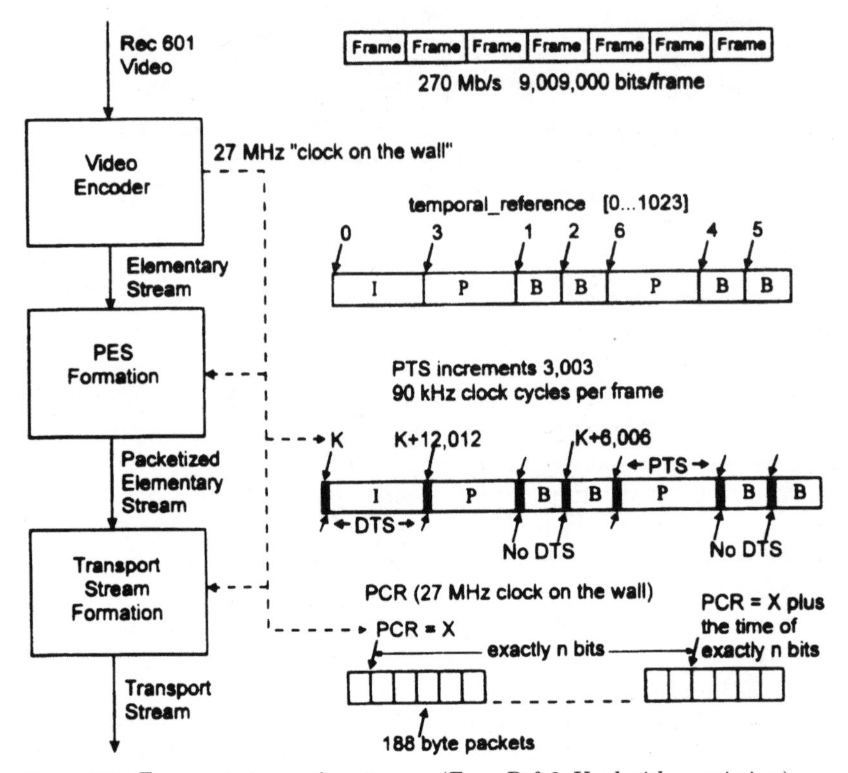

Figure 9.10 Transport stream time stamps. (*From Ref. 2. Used with permission.*)

maximum value of more than 24 h for programming convenience. The value may be calculated as

$$\text{PCR value (in seconds)} = \frac{\text{base}}{90\times 103} + \frac{\text{extension}}{27\times 106}$$

At the transport stream header where the PCR is inserted, the exact clock-on-the-wall time is given as the PCR value. The value for the subsequent PCR will be exactly the time required for the *e* bits that occur between PCR insertions.

9.3.4 Reference clock synchronization

The MPEG-2 system model assumes that the time for each transport stream packet to travel from the encoder to the decoder is constant. Therefore, a software-controlled PLL can be implemented to reconstitute the exact clock-on-the-wall frequency but with a relatively earlier time (due to transmission delay). A block diagram showing the concept of the decoder PLL is shown in Fig. 9.11. When the PCRs are

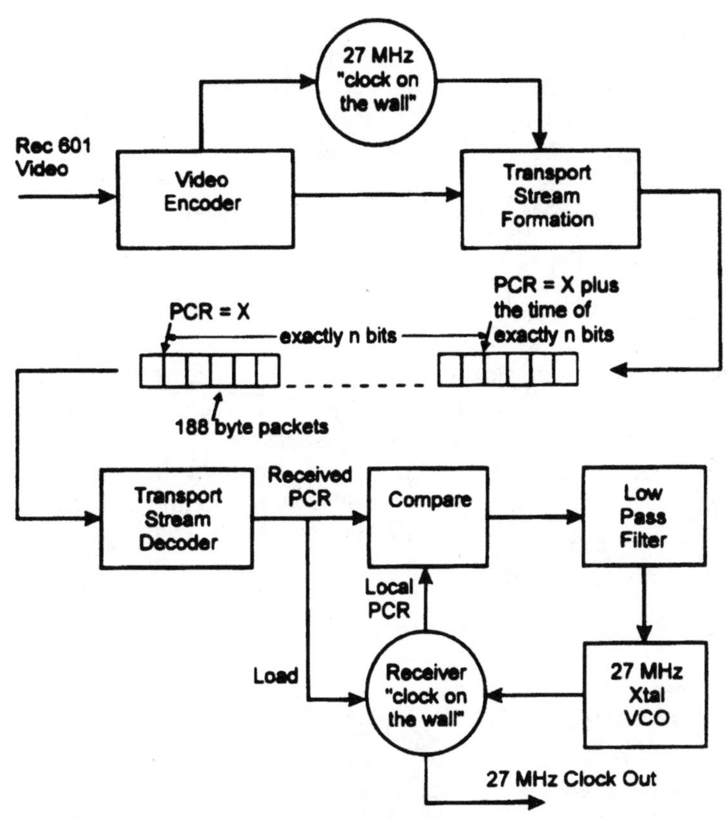

Figure 9.11 Encoder-to-decoder clock synchronization. (*From Ref. 2. Used with permission.*)

first received, or the continuity of the PCRs changes as a result of a program change, the "load" signal sets the local clock time. Then as each PCR is received, it is compared to a locally generated PCR and the difference, if any, is used to adjust the 27-MHz crystal-voltage-controlled oscillator. Tolerance of the 27-MHz system reference clock is required to be + 30 parts per million (ppm) or + 810 Hz, with a rate of change less than 0.075 Hz/s. In addition, the PCRs are required to have an accuracy of ± 500 ns, not including packet arrival time due to network jitter or other causes. This accuracy specification is due to the time placement of PCRs at low-transport-stream data rates.

Because the PCRs may occur as infrequently as 10 times per second, the PLL low-pass filter must be very narrow, such as 1 Hz. This creates both an advantage and a disadvantage. A properly designed PLL will eliminate any jitter above its nominal bandwidth. Considering bandwidths in the range of a few hertz, a stable TV signal can be reproduced. Unfortunately, a PLL with a low bandwidth will have a long lockup time (not good for channel surfing), so an adaptive loop time constant change is likely to be used.

Design of the receiver PLL is key to handling transmission channel timing variations (jitter), as well as the allowed accuracy of PCRs. For those familiar with TV genlock circuits, this is similar to developing a stable reference signal using only every third or fourth vertical sync pulse.

The fact that transport streams have a nominally constant bit rate may be an advantage in PLL design and can be used to evaluate the accuracy of the PCRs in received data. MPEG-2 is a receiver specification only, which allows continuing compatibility with new encoding methods as technology and computing power improve. For the transport stream a hypothetical decoder, the *transport stream system target decoder* (T-STD) is used to imply the specifications for the output of encoders. Although transport streams are not specifically defined as being so, all implementations have a constant bit rate. This practical result is emphasized by the MPEG-2 system standard.

9.4 References

1. Lakhani. G., "Video Compression Techniques and Standards," in J. C. Whitaker (ed.), *The Electronics Handbook,* Chap. 84, CRC Press, Boca Raton, Florida, 1996.
2. Fibush, D. K., "Timing and Synchronization Using MPEG-2 Transport Streams," *SMPTE J.,* SMPTE, New York, July 1996, pp. 395–400.
3. EBU N-10, "Parallel Interface for Analogue Component Video Signals."
4. ANSI/SMPTE 274M-1995, "Television—1920×1080 Scanning and Interface." (Originally published as SMPTE 247M, "Proposed Television—1920×1080 Scanning and Interface," *SMPTE J.,* vol. 103, October 1994, pp. 707–718).
5. ITU-R BT 601-5, "Studio Encoding Parameters of Digital Television for Standard 4:3 and Wide-Screen 16:9 Aspect Ratios."
6. ISO/IEC 13818-1995, "Information Technology—Generic Coding of Moving Pictures and Associated Audio Information: Part l-System, Part 2-Video, Part 3-Audio."
7. ITU-R Doc. 11-1/6E, "Information Paper, MPEG Digital Compression System," August 9, 1994.
8. IEE Colloquium, *MPEG-2, What It Is and What It Isn't,* IEE, Stevenage, Herts, U.K., January 24, 1995.

Chapter

10

Applications of Compression

10.1 Videoconferencing

As set forth in Chap. 1, transmitting and receiving audio and video for teleconferencing imposes some special requirements:

- *Symmetrical operation.* Both a dedicated encoder and a decoder are needed for full-duplex conferencing, and the cost of each system should be similar.
- *Real-time operational capabilities.* The communication must take place without perceptible delay. In the case of the H.261 specification, real time is less than 150 μs each way for processing (compression and decompression). The link may also have a delay. Even in the event of a news or sporting event broadcast, it may be acceptable for a delay of 5 s or much more. But in a conversational setting, this would result in considerable confusion.
- *Scalability.* Consideration of scalability is more important in videoconferencing than in some other applications. MPEG-1, for example, focused on the 1.2-Mbit data rate associated with compact discs, whereas videoconferencing could occur over LANs, WANs, POTS, ISDN, or T1 links.

10.1.1 H.261

CCITT Study Group XV (Transmission Systems and Equipment) began work on a videoconferencing standard in December of 1984 when they established a Specialists Group on Coding for Visual Telephony. The group originally tried to utilize compression in order

TABLE 10.1 Available Bandwidths for Different Data Links

Service	Bandwidth
POTS	28.8 kbits/s
ISDN	128 kbits/s
T1	1.54 Mbits/s
SDSL	2 Mbits/s

to realize data rates for ISDN applications of $m \times 384$ kbits/s, where m was an integer between 1 and 5. By 1989, they had broadened target data rates to $p \times 64$, with p an integer between 1 and 30. Table 10.1 lists the bandwidths for several common data links. Like the MPEG standards, H.261 specifies only the decoder, allowing for enhanced encoders as long as they produce a compliant stream.

Because of the constraint on processing delay mentioned previously, H.261 uses only the closest previous frame. Furthermore, because the reliability of the channel cannot be guaranteed, consideration is made of system robustness under adverse conditions. Lastly, the parameters and coding structures are optimized for lower-bit-rate applications at the expense of performance with higher bit rates. Multiples of 64 kbits/s up to 2 Mbits/s do not produce a perfect subset type of scalability.

As shown in Fig. 10.1, the H.261 bit stream is comprised of four layers:

- The picture layer
- The GOB layer
- The macroblock (MB) layer
- The block layer

Figure 10.2 illustrates the block structure. Committee involvement is apparent in the combinations of spatial and temporal resolutions supported. H.261 specifies a luminance spatial resolution of CIF 352 pixels per line by 288 lines per frame (which is conveniently one quarter of a PAL frame) and a 29.97 frame-per-second rate (which is the same as that of NTSC). All standard codecs must be able to operate with QCIF, while CIF is optional.

A compressed bit stream begins with the picture layer data header, which includes the following elements:

- *Picture start code (PSC).* 20 bits
- *Temporal reference (TR).* 5-bit input frame buffer
- *Type information (PTYPE).* For example, CIF/QCIF selection
- *User-inserted bits*

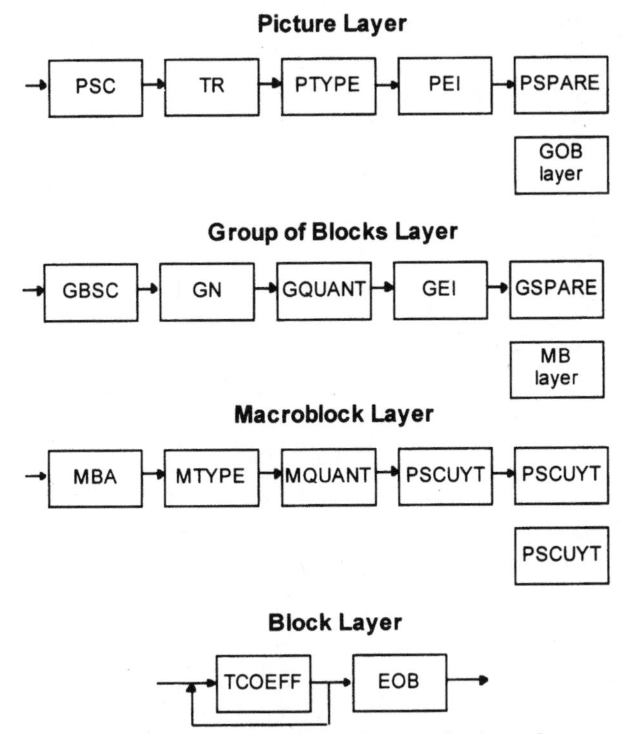

Figure 10.1 Syntax diagram of the H.261 multiplex coder.

The GOB layer data header includes the following:

- *GOB start code.* 16-bit pattern
- *Group number* (*GN*). 4-bit GOB address
- *Quantizer information* (*GQUANT*). Quantizer step-size normalized to lie in the range 1 to 31
- *User-inserted bits*
- *Macroblock layer data*
- *Macroblock address* (*MBA*). VLC location relative to the previously coded MB
- *Type information* (*MTYPE*). 10 types in total
- *Quantizer* (*MQUANT*). Normalized quantizer step size
- *Motion vector data* (*MVD*). Differential displacement
- *Coded block pattern* (*CBP*). Coded block location indicator

A simplified block diagram of an H.261 encoder is given in Fig. 10.3.

Blocks in a MacroBlock

1	2
3	4

5

6

MacroBlocks in a GOB

1	2	3	4	5	6	7	8	9	10	11
12	13	14	15	16	17	18	19	20	21	22
23	24	25	26	27	28	29	30	31	32	33

GOBs in a picture

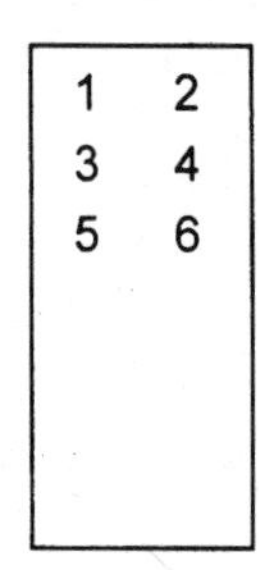

Figure 10.2 H.261 coding scheme.

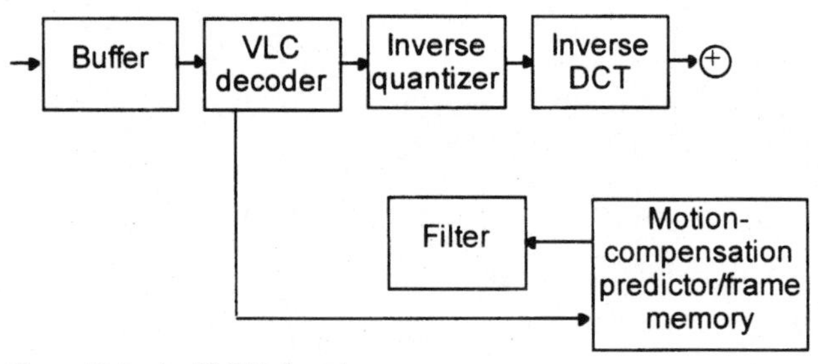

Figure 10.3 An H.261 decoder.

10.1.2 H.261 encoding

An encoder can be thought of as having two basic types of blocks:

- *Basic coding operation units.* These include the motion estimator, quantizer, discrete cosine transform, and variable-word-length encoder.
- *Coding parameter decision units.* These select the parameter values for the basic operation units, including motion vectors, quantization step size, and the picture frame rate.

Because H.261 does not specify an encoder, some design elements of the encoder, such as the parameter decision unit, are open to the designer.

10.2 Digital Video Transmission

The discussion so far has centered on processing the digitized signals. However, it is not enough for these data bits to just sit in one place—they must travel to be seen and heard. When these signals are brought into the home, *video on demand* (VOD) and other advanced services can be accessed via PC or television.

Many people spend much time and effort debating whether PCs or TVs will be the dominant multimedia center in the home. That discussion really centers on the long term and, as an economist once pointed out, "in the long run we're all dead." In the short run, while we are all alive, TVs and PCs will both be in the home and used for receiving compressed video and audio. Mass deployment should develop along the time line given in Table 10.2.

10.2.1 TV reception model

The traditional methods of connecting TVs to the outside world are, of course, a standard terrestrial antennae, satellite dish and set-top box (STB) combination, or a cable STB. For PCs, telephone lines are the traditional connection method.

The broadcast model for data delivery is similar to that used by most radio frequency (RF) transmissions today; that is, specific RF channels in the very high frequency (VHF), ultrahigh frequency (UHF), or microwave bands are allocated to the signal(s) being transmitted. This is referred to as *frequency-division multiplexing* (FDM). The only differences center around the fact that the signals sent are digital. The first difference is that binary data can be easily *time-division–multiplexed* (TDM). Using TDM, multiple transmissions can be combined onto one RF carrier. Another difference is that, even if the data rates increase, the RF spectrum can be held constant using advanced vector modulation techniques. Furthermore, using digital

TABLE 10.2 Video Service Introduction Timetable

Service	Year
Digital satellite TV	1994
Digital cable TV	1995/6
PC cable modem	1996/7
Digital TV/HDTV	1998/9

forward error correction (FEC), the signals can be received at a lower *carrier-to-noise* ratio (CNR) than possible with analog signals. This means with smaller satellite receiving dish antennae and with cable that there is less chance of interference with analog channels that may be present.

In the new systems, TDM, vector modulation, and FEC and FDM are combined. This combination allows for hundreds of digital TV channels plus music and data services, where only up to 80 to 120 channels of analog TV could exist before.

At the transmit side, the digital signal for a particular channel is then modulated onto a carrier in the proper frequency band at the transmitter. At the receiver, the signal is converted from its particular channel down to an intermediate frequency (IF) for processing. This IF signal is demodulated and becomes raw data. These data have an associated *bit error rate* (BER). After processing through the FEC decoding block, the BER reduces by orders of magnitude and it essentially error free.

Tuning systems facilitate the conversion of any of the RF channels to the IF. Converters for TV come in either single-conversion (low-cost) or dual-conversion (low-distortion) varieties. Typically, dual conversion is used in a cable TV set-top box, because of its frequency response and distortion improvement. However, because of the potential for cost savings, people have been exploring the use of single-conversion tuners for digital receivers.

The other important issue with digital tuners is phase noise. This type of noise degrades the ability of the digital receiver to distinguish between different data points. Low-phase noise tuners will be essential in digital receivers.

The demodulator converts the IF signal to baseband. Most modern demodulators use digital-processing techniques for high performance, operational versatility, and cost considerations.

Error-correction techniques are used to reduce the receiver output data BER by several orders of magnitude. Typical FEC output BER will be approximately 1×10^{-12}. The FEC techniques include Viterbi (Trellis) encoding and Reed-Solomon encoding. In the case of satellite transmissions and some cable standards, these two techniques are concatenated to give optimum results.

10.2.1.1 Satellite delivery. In satellite systems the signal is up-converted to microwave frequencies (typically 2 to 12 GHz). On the receiver end, a low-noise down converter transforms the signal into the tuner's input range (typically 950 to 1750 MHz). The tuner IF output is typically anywhere from 70 to 610 MHz. This signal is then fed to a digital demodulator. Although not as spectrally efficient as other higher-order m-ary PSK methods, QPSK modulation is used in satellite TV to keep the noise and reflection immunity high.

10.2.1.2 Cable delivery. In cable systems the signal is up-converted to VHF and UHF frequencies (50 to 750 MHz). On the receiver end, a tuner processes the signal and produces an IF output of typically 45 MHz. This signal is then fed to a digital demodulator. Although not as robust as QPSK methods, 64 and 256 QAM modulation is used on cable systems to keep spectral efficiency high.

10.2.1.3 Wireless cable. Not the first oxymoron to be used in this book, wireless cable is a method proposed by many equipment manufacturers and potential operators because of its potential for easy, low-cost implementation in new market areas. Wireless cable franchises are operated under the Multichannel Multipoint Distribution Service (MMDS) guidelines established by the FCC in the United States.

The MMDS is a point-to-multipoint broadcast of signals using terrestrial microwave transmission. Transmitting antennae are omnidirectional or multidirectional and are placed within the line of sight of the target receivers. Other than the input frequency range and the potential for interference from other transmitters on the same frequency (when configured in a cellular configuration), a wireless cable box can look very similar to a cable box. Typically, interactivity is limited to a phone-line return path, as in a satellite, and the modulation type is limited to 64 QAM.

10.3 CD Applications

During the 1970s a number of companies studied digital audio systems as modern alternatives to the vinyl disc. There were a number of proposed systems, but with the format wars of videotape and quadraphonic sound fresh in their minds, Philips and Sony agreed to jointly propose a variation of the original Philips compact disc (CD) standard, which was based on optical technology. The success of this format in the audio industry in the 1980s has provided an economy of scale for other applications, and it holds the promise of extending the product life into the next century by adding compression and higher density.

Because the spectrum of possible data storage uses could not be anticipated when the CD was first popularized for audio applications, a rainbow of reference books has subsequently been published by Philips with various partners that describe other applications.

TABLE 10.3 CD Specifications

	Read Only	
Red book	Digital audio (CDDA)	1982
Yellow book	Basic data application	1984
Green book	CD interactive	1987
(CD-ROM-XA	Extended architecture	1989)
White book	CD-video	1993
Enhanced CD	CD-plus	1995
	Recordable and Erasable	
Orange book	Part 1: (magneto-optic & CD-R)	1991
	Part 2: CD-recordable	1991
	Part 3: CD-erasable	Pending

10.3.1 CD physical specifications

All extant CDs today are mechanically compatible. They are all the same thickness, 1.2 mm, and the same diameter, 12 cm (just under 5 in) or less (commonly 8 cm). All CDs store the raw data in the same way, such that the 1s and 0s can be read as such by all players. Further information on the basic CD specification can be found in Table 10.3.

Data are stored on the disc as a line of elongated pits along a spiral track that begins near the center of the disc and proceeds toward the outer edge. The main concept to note here is the way in which the bits are denoted on the disc. The pits have lower reflectivity compared to the land areas of the disc, enabling the laser beam to detect the pits, stay on track, and decipher the stored data using digital signal processing. A change from a land to a pit or a pit to a land is represented by a logical 1. A no change (continuous land or pit) is represented by a logical 0. The decision is made by comparison of the status of the output signal against a clock of 4.3218 MHz locked to the received data. This technique enables the reading of data from a stamped disc by use of a scanning laser beam. The reflective layer containing the information is sandwiched between two polycarbonate layers. The read-out side of the sandwich is thicker than the back side containing the label, making it more sensitive to mechanical mistreatment, contrary to popular belief.

10.3.2 CD system standards

The first of the standards books, which describes Compact Disc Digital Audio (CD-DA), was published in 1982 with a red cover. The so-called red book sets forth the physical dimensions, basic data formats, and the audio application. Over the years Philips and Sony, in

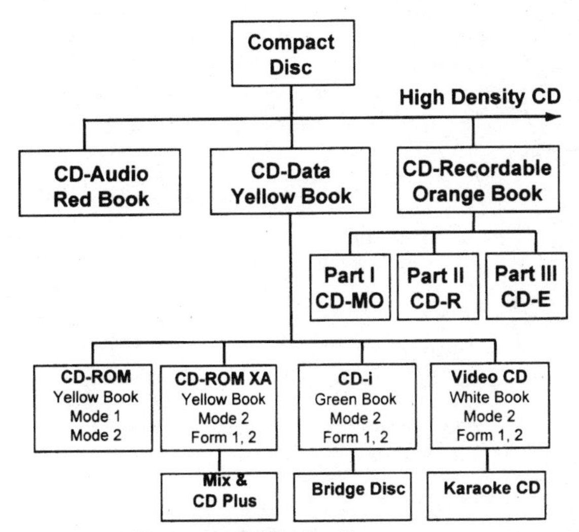

Figure 10.4 The compact disc family tree.

conjunction with other companies, have published several other specifications to expand the capability of the CD into other applications. Table 10.3 lists the main standards released to date. The compact disc family tree is given in Fig. 10.4.

The success of CD-DA has made it possible to store large amounts of data at very low costs. Today the cost of storing data on a CD-ROM disc is less than US $ 0.002 per megabyte. Following the release of the red book for digital audio, the yellow book was introduced in 1984, defining the basic format for data storage applications. All other data formats are based on the yellow book.

The green book for CD Interactive is a variation of the yellow book with the inclusion of a specific operating system called CD-RTOS (real-time operating system) as well as graphics and ADPCM audio compression formats. The CD-ROM XA sector format and the sub-header fields were defined in conjunction with Microsoft, photo CD was defined in conjunction with Eastman Kodak, and Karaoke CD—including full motion video—was defined in conjunction with Japan Victor Company (JVC), supported by Sony and Matsushita. Karaoke CD was later renamed to be Video CD, and the standard is defined in the so-called white book. The white book defines the format for storing full motion compressed video using ISO MPEG-1 audio and video compression. Enhanced CD, also referred to as the CD Plus format, was defined to enable CD-ROM discs to be combined with CD-DA that can be played on CD audio players. The so-called orange book defines the requirements for recordable and erasable CD systems.

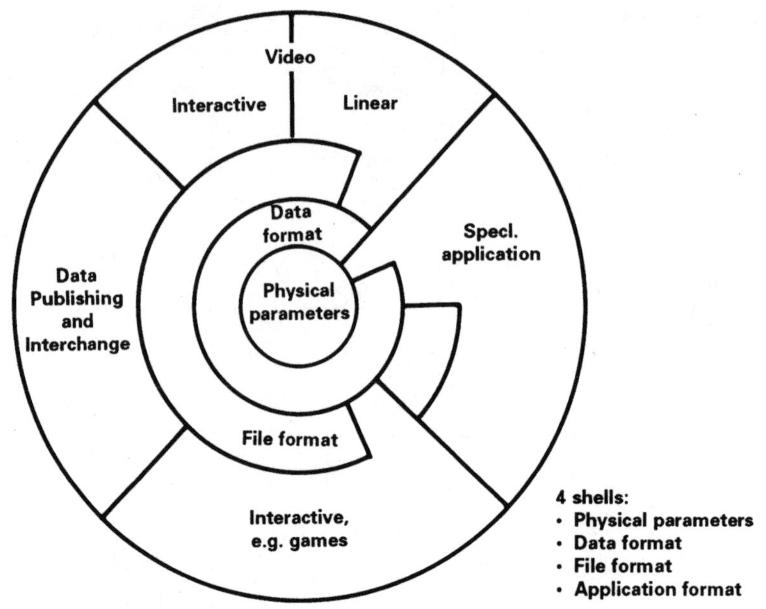

Figure 10.5 Model of the physical, data format, file, and application shells.

It is important to note that the standards books from Philips define only the physical parameters and the data formats. There are other layers of standards involved in generating a disc for an application, as shown in Fig. 10.5. In addition to the standard CDs there are also proprietary disc formats (e.g., in games platforms such as CDTV, Sega, Play Station from Sony, and 3DO) used in specific applications that may or may not be readable on standard audio or CD-ROM readers. When storing computer files (data) on a compact disc, it is necessary to follow a file structure with a hierarchical directory configuration, so that the system can locate the data quickly and efficiently, similar to the way IBM-compatible PC systems use the Microsoft disc operating system (MS DOS) and Apple Macintosh machines use the hierarchical file system (HFS) for storing files on hard disks and floppy disks.

10.3.3 Red book: audio CD

Virtually no compression is used in the basic audio CD application. Because the audio is digital, it has been sampled and quantized, which we identified as a form of data reduction in Chap. 1. In spite of all the wonderful compression techniques we have learned in this book, it was correct to choose to not use digital compression. At the time of product introduction in 1982, the cost of performing the digital signal processing in integrated circuits would have been prohibitive for a consumer product. Sixteen-bit audio samples (yielding 96-dB dynamic range) for each channel, left and right, stream out at 44.1 kHz per channel. These samples are converted to an analog signal using a digital-to-analog converter. Different pieces of audio data can be stored on the disc as separate tracks and indexes within a track, 99 of each being allowed in the red book specification. The maximum time of about 74 min of audio recording is equivalent to about 750 Mbytes of data.

All CD systems, audio and data, must follow the red book data format as the basic encoding scheme. The process begins by sampling the two analog audio channels (left and right) with 16-bit accuracy (bits per audio sample) at the rate of 44,100 samples per second per channel. Blocks of six samples from each channel (12 in all) are taken, and then each 16-bit sample is split into two 8-bit bytes, referred to as *symbols,* resulting in 24 symbols. For error correction purposes two sets of parity are generated, each with four symbols. The first four parity symbols, referred to as *C1 parity,* are generated operating on a block of 24 symbols as mentioned before but with additional delays and scrambling among the blocks for further enhancement of error correction. The exact details are described in the red book and are beyond the scope of this publication. Similarly, four more parity symbols, referred to as *C2 parity,* are generated using 28 symbols from the C1 stage (24 data and 4 of C1 parity), along with the resulting in 32 parity symbols in total. As in the C1 stage, the C2 stage also includes symbol delays and scrambling. After generation of the parity symbols, one more byte containing control information commonly referred to as *subcode* data is added to each of these frames. Each bit in the subcode byte is a data channel on its own referred to as *Q, R, S, T, U, V, W* channels. The *Q* channel contains the most commonly utilized information such as minutes, seconds, index number, and other information relating to the disc. The *R* to *W* channels are used to record CD graphics data, and some remain reserved for future use.

After adding the subcode byte, the 8-bit data symbols are modulated using *eight-to-fourteen modulation* (EFM). This is done by means of code conversion, in which 8-bit bytes are allocated 14-bit words based on a conversion table. Using EFM modulation guarantees the continued reversal of data bits on the disc irrespective of the content of the data (e.g., even if all the data are zero). After EFM conversion the 14-bit words are serialized and three more bits, referred to as *merging bits,* are added between each of the 14-bit words, merging them together into a string of 564 bits. The 564-bit sector consists of 33 words of 17 bits each, plus three merging bits at the beginning and an additional three merging bits for closing. Finally, a 24-bit synchronization pattern is added to the 564-bit string, making a total of 588 bits, which is referred to as an EFM frame. Continuous streams of EFM frames containing 588 bits each are written onto the disc.

The three merging bits added to every EFM word serve two purposes. The first is to prevent violation of EFM modulation rules when serializing the words into a stream of bits. The second function is to move the low-frequency content of the data stream when reading back from the disc to a higher domain in the spectrum, so that the servo-control system can derive the information it needs to correctly position the laser. Figure 10.6 illustrates the data encoding process for a CD.

10.3.4 Red book playback

After retrieving the data from the disc, the reverse of the encoding process, de-interleaving and de-scrambling, is done while correcting any error that may have occurred in the process. At the end of the decoding routine, the original 24 symbols from the 6 audio samples of 16 bits each are recovered. On a single-speed system, these 24-byte frames are produced at the rate of 7350 Hz, resulting in the original sample rate of 6×7350 = 44,100 Hz. The combination of 98 of these audio frames forms an audio sector containing 2352 bytes (24×98) of audio data. Hence, the audio sectors are delivered at the rate of 75 Hz. The subcode data byte in each sector is separated into 8 data bit channels generating *Q* through *W* data. Audio players utilize only the *Q* channel data to determine information such as track number, index numbers, and time on the disc for display on the front panel. The *R* through *W* channels are mainly used for storing CD graphics (CD-G). Only the players equipped to decode these channels can utilize the information. The most common application is to display graphics pic-

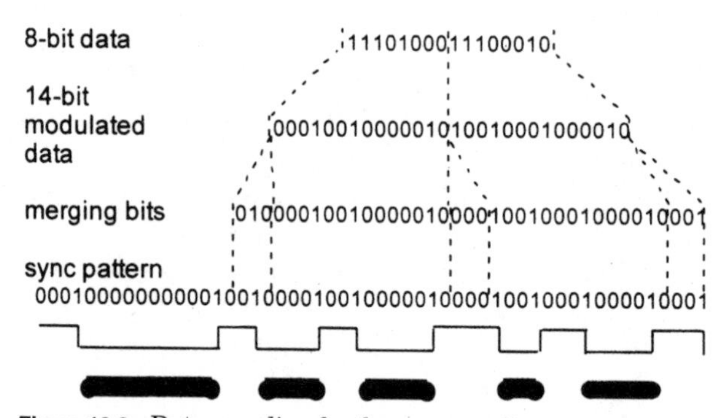

Figure 10.6 Data encoding for the compact disc.

tures, lyrics, information on the performer, etc., while playing audio. In order to display this information the player is connected to a display of some sort. Some CD-ROM players also offer this capability.

In the case of non–CD-DA applications, the 16-bit bytes represent data instead of audio samples. Also an optional error-correction stage, called the third layer (C3), and additional blocking of sectors with a sync pattern and a headers is used. Further details are given later in this chapter.

Ordinarily a red book disc contains only one *session*. A session consists of a lead in, a program area, and a lead out. As we will see later, it is possible for a disc to contain more than one session. Only players with multisession capability can read sessions beyond the first, such as some CD-ROM players and photo CD players. Figure 10.7 shows a typical CD-DA disc with a single session.

10.3.5 Yellow book: CD-ROM

The yellow book was introduced to enable data storage by extending the red book format, commonly known as the CD-ROM standard. The yellow book is built on the red book format by defining a sync pattern at the beginning of each sector followed by a header to indicate information about the data in the sector. The 12-byte sync patterns identify the beginning of the sector; a 4-byte header follows, where each byte represents minutes, seconds, and blocks (frames) relating to the absolute time as recorded in the subcode channel. These data are followed by a mode byte to indicate the operating mode of the disc. The header is followed by the main-user data area and an optional C3 third-layer error-correction code (remember that C1 and C2 are the first two layers of error correction in CD-DA).

There are two modes in the CD-ROM sector format. Mode 1 includes the third-layer error-correction capability. A sector in mode 1

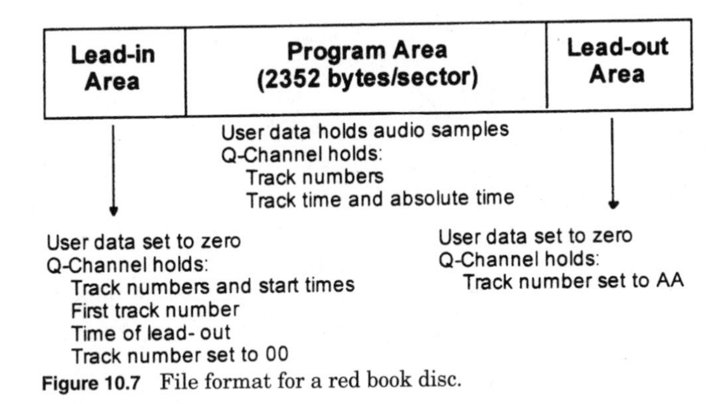

Figure 10.7 File format for a red book disc.

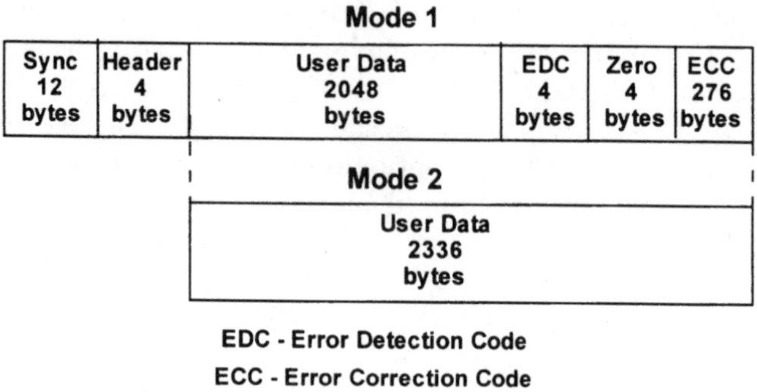

Mode 1

Sync 12 bytes	Header 4 bytes	User Data 2048 bytes	EDC 4 bytes	Zero 4 bytes	ECC 276 bytes

Mode 2

User Data 2336 bytes

EDC - Error Detection Code

ECC - Error Correction Code

Figure 10.8 CD-ROM sector layout according to the yellow book.

has 2048 bytes of raw data, whereas a sector in mode 2 utilizes the bytes used by the C3 code in mode 1 for raw data, increasing the raw data capacity to 2336 bytes. Figure 10.8 shows the relationship between the sectors of modes 1 and 2 of the yellow book CD-ROM format. CD-ROM XA (extended architecture) is an extension of CD-ROM mode 2.

10.3.6 ISO 9660 format

As a result of deliberations by the computer industry in the High Sierra Hotel & Casino in Nevada (hence originally referred to as the *High Sierra* file format), a file format for CD-ROM systems was defined. This proposal was later adopted by the ISO with few refinements and is referred to as ISO 9660. In order for an IBM-compatible computer to be able to read CD-ROM discs with the ISO 9660 file structure, additional software is required to interface with the operating system. This was provided by Microsoft and referred to as Microsoft CD extension (MSCDEX.EXE), which interfaces the device driver provided by the CD-ROM drive supplier and the operating system. ISO 9660 is a universal file format for CD-ROM, enabling CD-ROM discs to be read on a number of different platforms, such as Apple Macintosh and UNIX. Additional extensions to ISO 9660 may be added to enhance its capabilities on some of the platforms; for example, Apple Macintosh has extensions using the system user field of ISO 9660 to maintain some of its HFS file systems benefits. On a CD-ROM disc there is at least one file system. One session contains only one file system, but a single file set may be extended over more than one session. Figure 10.9 shows the layout of a yellow book disc with the file structure information. The ISO 9660 set is characterized by the following parts:

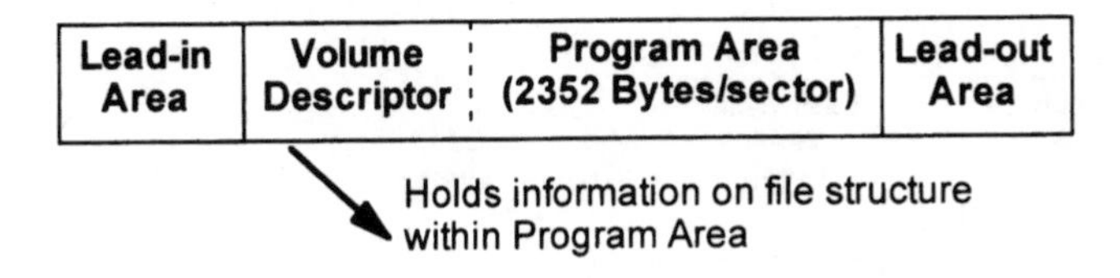

File structure conforms to ISO9660 formats

Figure 10.9 A yellow book disc file structure.

- *Primary volume descriptor* (*PVD*). This is stored in the sixteenth sector of a session. This is the entry point in the medium to the file system. It refers to the location of the root directory and the path table.
- *Path table* (*PT*). This contains the locations of the directory files and is used for direct access of the files.
- *Root directory* (*RD*). This is the main directory, and hence no other directory refers to it.
- *Directory file* (*DF*). This contains the locations of the subdirectories and data files. Subdirectory information is used to descend through the directory tree in search of data file locations.
- *Data files*. These contain the actual data stored on the disc.

If there are multiple sessions on a disc, the PVD, PT, and DF are allowed to refer to the data files of the previous session. However, the addressing of data sectors are absolute and continue over the whole disc throughout all sessions. This enables extension of one ISO set over more than one session.

It is important to note that a multisession disc with disjunct ISO sets (with no reference to the previous session) will appear to MS-DOS as two drive letters, e.g., E:\ & F:\, the same way as a partitioned hard disk.

10.3.7 Green book (CD-I) and yellow book CD-ROM-XA

Compact Disc Interactive (CD-I) was introduced by Philips in 1987. In addition to the disc layout and sector format, the CD-I standard defines its own operating system called *CD real-time operating system* (CDRTOS), based on OS9 from Microware, enabling stand-alone application of the system by connecting a CD-I player to a standard TV.

CD-ROM mode 2 was extended to include a subheader after the header, as described previously under the yellow book section. The

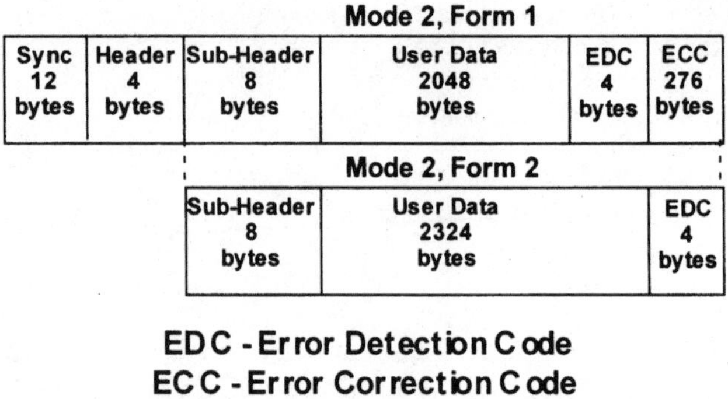

Mode 2, Form 1

Sync 12 bytes	Header 4 bytes	Sub-Header 8 bytes	User Data 2048 bytes	EDC 4 bytes	ECC 276 bytes

Mode 2, Form 2

Sub-Header 8 bytes	User Data 2324 bytes	EDC 4 bytes

EDC - Error Detection Code
ECC - Error Correction Code

Figure 10.10 Yellow book CD-ROM-XA and green book CD-I sector layout.

use of the subheader enables multiple data streams and files to be interleaved in the same track. CD-ROM-XA was introduced in 1989 by Philips, Microsoft, and Sony. The CD-ROM-XA and CD-I use the sector layout shown in Fig. 10.10. The 4-byte subheader contains the file number byte, channel number byte, a flag byte, and a coding information byte. The file number enables the interleaving of different logical files, up to 255 of them, into the single data stream. The channel number enables multiple (32) streams of audio, video, and data to be combined into a single data stream by interleaving sectors (only 16 channels of audio allowed). An example of a practical application is a video stream with audio streams of multiple languages for the user to select from, with interactively overlaying animation sequences. Mode 2 also has the capability to utilize third-layer error correction. When error correction is to be used, the form 1 sector layout is used and form 2 is used in all other cases.

In order to effectively achieve multiple streams within the available bandwidth of the CD-ROM data stream, compression of audio and video becomes a necessity. The current trend in using higher speeds of the CD-ROM disc will enable application developers to define the minimum required speed, for example, 4× (four times the standard data rate of 150 kbytes/s), and in so doing provide higher data rates per individual stream at the expense of shorter play time. The introduction of the high-density CD will also push the data rate and capacity by over 10 times the current CD standard. The availability of higher data rates and higher capacities is not expected to diminish the demand for higher compression algorithms. Continuing demand for improved performance of video, audio, and higher interactivity will utilize all the improvements expected in the coming years.

TABLE 10.4 CD-I Audio Formats

Audio stream	Maximum play time	Number of bits per sample	Sample rate, kHz	Megabytes per 1 min of stereo audio
CD-DA	1 hr, 14 min stereo	16	44.1	10.09
Level A	2½ hr stereo			
5 hr mono	8	37.8	4.33	
Level B	5 hr Stereo			
10 hr mono	4	37.8	2.16	
Level C	10 hr stereo			
20 hr mono	4	18.9	1.08	

10.3.8 ADPCM audio compression in CD-I

The audio compression defined initially in CD-I and later adopted in CD-ROM-XA uses adaptive pulse-code modulation. The CD-I system defines three levels of audio compression, referred to as levels *A, B,* and *C,* of which CD-ROM-XA uses only levels *B* and *C.* (See Table 10.4.)

10.3.9 CDI-ready disc

This is a red book disc with additional CD-I features but is not visible to a standard CD-DA player. Red-book-format discs have a minimum of a 2-s pause period in track 1 during which the index number remains zero. Index numbers 1 through 99 are allowed during any track number other than zero to identify specific parts of music. Audio recordings start after the pause period; and hence audio-only players (CD-DA players) skip over this part of the track. In a CD-I ready disc this pause period is extended and CD-I information is recorded. When playing such a disc on a CD-I player, it can read the hidden information and provide additional features along with the music, such as, for example, lyrics, information about the title, or information about the artist. The hidden information may even contain compressed audio, graphics, or data. The CD-I information is loaded into the system before playing the red book audio, giving the pretense that the additional data are being read at the same time as the audio.

A similar technique is sometimes used to generate CD-ROM–ready discs. These discs use track 0, referred to as the *pregap* area. Normally, the main data are zero during zero track. In CD-ROM–ready discs, the pregap is extended to store CD-ROM data. Such a disc can be read on a CD-DA player, but when used on a CD-ROM player the additional information can be utilized to offer additional features. The reason for generating this format is to add data to a disc with the audio tracks in such a way that the installed base of CD-DA players can read the audio tracks; the data can be extracted by playing the disc on a computer using a CD-ROM drive. This requirement was later fulfilled with the introduction of the enhanced CD standard in 1995.

10.3.10 Enhanced CD (CD Plus)

Similar to CD-I ready discs, it is possible for CD Plus discs to be used to down load additional information about the audio tracks into a computer and use that information in real time to control the playback of the audio track(s). The format may be simply used to include CD-ROM data along with audio CD. Such discs are also used for promotional applications where the user can play the disc on a home CD player and simply listen to the audio tracks and also use the same disc on a computer with a CD-ROM reader for interactive ROM features. The format defines two sessions, the first being a CD-DA session followed by a data session.

10.3.11 CD-Bridge Disc

The CD-bridge disc primarily contains CD-ROM or CD-ROM-XA tracks with additional information to permit the user to play the same disc on more than one platform (for example, to use a CD-I disc on a CD-ROM, photo CD, 3DO, or Karaoke). Standardization accommodates this by allocating two separate areas for storing the primary volume descriptor, as described earlier under ISO 9660. In the case of CD-I, the PVD is located at 00 minutes 02 seconds and 16 frames with a zero offset (Fig. 10.11). For CD-ROM-XA, the volume descriptor is stored with an offset of 1024. Bridge disc tracks are included in the table of contents (TOC) as CD-ROM-XA, whereas in CD-I discs the track information is not included in the TOC to prevent CD-DA players from playing the CD-I data tracks and possibly damaging the speakers.

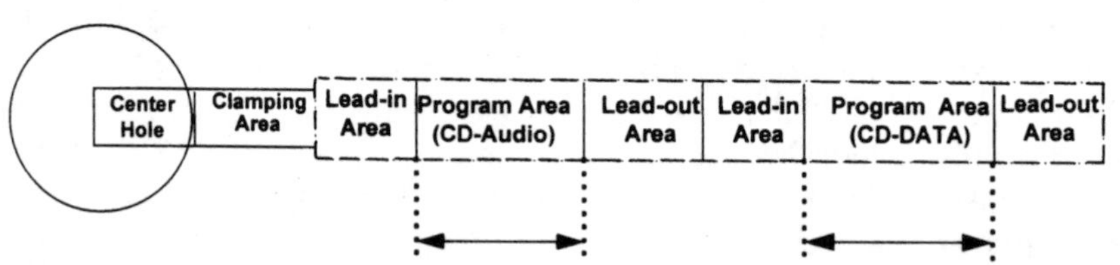

Figure 10.11 CD-Plus disc layout.

TABLE 10.5 Photo CD Formats

	Format	Number of pictures per disc	Application
1.	Master	100	Color prints at the photofinisher
2.	Portfolio	800	Video resolution combined with sound for presentations
3.	Catalog	6000	Catalog applications with programmed branching capability
4.	Pro	Varies	For professional 35 mm, 120, 4×5, and 8×10 in storage
5.	Medical	Varies	X ray, CT/MR scanner image storage

10.3.12 Photo CD

As the name suggests, the photo CD was developed to store photographs electronically. The user can select from a choice of resolutions for storing images, as described in Table 10.5. CD-R is a write-once medium, and hence a method had to be developed to keep adding new pictures when necessary. The *multisession* standard was originally defined for use with CD recordable discs for photo CD applications. Every time a set of photos is stored, a new session is generated. Discs with completed sessions can be read on photo-CD–compatible players. To read multisession discs, CD-ROM drives must be designed for this capability (drives sold before the end of 1994 may not have this capability).

Photo CDs essentially follow the CD bridge format to enable reading them on CD-I and CD-ROM-XA drives (also 3DO). The photo CD can be mastered and replicated as an ordinary CD. The photo CD system defines five picture resolution formats. The storage capacity occupied by an image can vary from 48 kbytes per image of 128×192 resolution to as much as 18,438 kbytes for a 2048×3072 high-resolution image. The photo CD uses two image compression techniques. Chrominance is subsampled relative to luminance by one-half, both vertically and horizontally. Data are compressed by Huffman coding of residual components. Audio information can be interleaved, and ADPCM compressed data are used as in CD-ROM-XA. CD-ROM and CD-I discs may include photo CD format pictures for storing still images.

10.3.13 CD-I full motion and video CD

Full motion video (FMV), originally used in CD-I and later in video CD, uses the MPEG-1 standard as described in Chap. 9 for compressing audio and video data. The video compression ratio realized is about 50:1. The video CD standard is defined in the white book. As

TABLE 10.6 Video-CD Parameters

Resolution	352 h×240v, 29.97 Hz (NTSC); 352 h×240v, 23.976 Hz (film); 352 h×288v, 25 Hz (PAL)
Pixel aspect ratio	1.0950 when resolution is 352×240, 0.9157 when resolution is 352×288
Compressed bit rate	Maximum 1151929.1 bits/s Audio compressing ISO MPEG (3-11172) layer 2
Audio modes	Stereo, dual mono, intensity stereo
Audio sampling rate	44.1 kHz
Audio emphasis	Off or 50/15 μs
Compressed bit rate	224 kbits/s

the bandwidth available for the data stream is more important than the correctability of occasional errors, the compressed audio and video use the mode 2, form 2 sector format, without the error correction. Audio and video sectors are interleaved in such a way as to maintain the proper data flow rates. Details of the video CD format are given in Table 10.6. In the FMV/video CD format, 74 min of full motion video with audio can be stored on one disc. Distribution of movies and karaoke are the two most popular applications.

It is important to note that CD-I full motion discs can only be played on CD-I players with the additional hardware for decompression of MPEG-1 and the capability of reading the interactive file structure. Video CDs can be played on CD-I players with FMV capability, standalone video CD players connected to a TV, CD karaoke players, and on computer platforms with a CD-ROM drive and hardware or software to decode the MPEG data. Video and karaoke CDs utilize the ISO 9660 logical file format to store additional information (user-oriented information). In addition to full motion MPEG video and audio, the white book also defines a MPEG-SP (still picture) format.

10.3.14 Recordable CDs

Recordable CDs differ from pressed CDs in a number of ways. The construction of the disc includes a transparent plastic substrate that is the same as in a conventional disc, but it has a spiral track. On top of the substrate is a greenish translucent layer, and on top of that is a gold layer protected by another plastic layer. On recordable CDs there are two special areas located before the lead-in area of the disc. These areas are for interim storage of information required in generating a CD.

One area is referred to as the *program memory area* (PMA), which stores the start and stop times of each track. Using this information the TOC is generated during the finalization of the written disc, enabling playback drives to locate track positions as specified in the

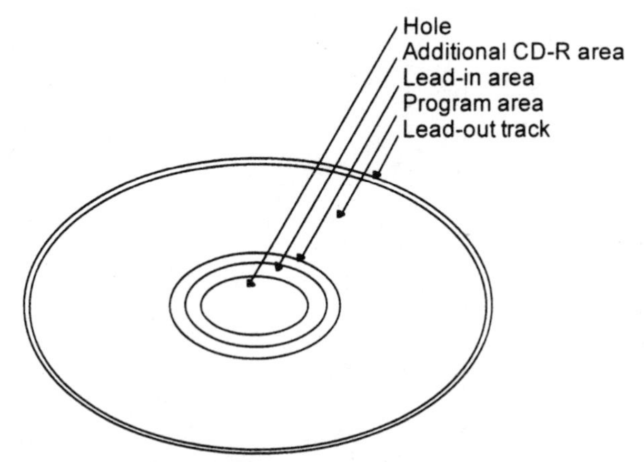

Figure 10.12 Layout of a CD-R disc.

red book. A CD-R disc cannot be read on a standard CD-ROM or CD-DA drive before finalization of at least one session because the TOC does not exist. The PMA area is physically situated to the interior of the lead-in area of the disc and hence is not normally accessible by CD-ROM and CD-DA players (Fig. 10.12).

The second special area is referred to as the *power calibration area* (CPA). This area is used for trial recordings to determine the optimum laser power required to write before data are written to the disc. Depending on the type of drive, the type of disc, and the environmental conditions such as humidity and temperature, the optimum laser power value will vary. The number of calibrations using the PCA is limited to 99 times, according to the orange book. Normally a disc is calibrated every time it is inserted in the drive for writing. However, different manufacturers use different techniques to maximize the use of PCA. Some of these techniques include using shorter writes to conserve the area or storing the disc ID and the calibration information on the host computer, thus preventing the need for recalibration every time the same disc is inserted.

10.3.15 CD-video

CD-video is perhaps the shortest lived of the various and sundry formats and is not to be confused with the now-popular video-CD. A CD-video disc started as a digital audio disc and ended with some analog laser disc video. Given the relatively small installed base of laser disc players and the subsequent stampede to all-digital formats, this scheme never attained critical mass. Principal specifications for the CD-video system are given in Table 10.6.

TABLE 10.7 A Comparison of CD-ROM and MMCD

	CD-ROM	MMCD
Disc diameter, mm	120/80	120/80
Disc thickness, mm	1.2	1.2
Program area ID-OD	25–58	23–58
Wavelength, nm	780	635
Numerical aperture	0.45	0.52
Track pitch, μm	1.6	0.84
Shortest pit/land, μm	0.83	0.45
Pit length increment, μm	0.28	0.15
Physical density	1	3.5
Reference speed, m/s	1.2	4
Reflectivity	High	High or low
Number of layers	1	1 or 2
Data capacity, GB	Mode 1: 153.6 Mode 2: 176.4	1400
Reference user data rate, kbits/s	Mode 1: 153.6 Mode 2: 176.4	1400
Burst error correction, B	500	2200
Raw disc byte error rate	2×10^{-3}	2×10^{-2}
Sector size (user)	2048	2048
Video format	MPEG-1	MPEG-2
Video data rate Mbits/s	1.15	1–10, average 3

10.3.16 MMCD

The next generation of CDs is the *high-density compact disc* (HDCD). Because it is specified in consideration of multimedia applications, it is commonly referred to as *multimedia compact disc* (MMCD). Application design goals included maintaining backward compatibility with existing CDs (i.e., new MMCD players can play old CD discs). A single-layered disc provides 135 min worth of MPEG-2 data. In this application it is called a digital video disc. Table 10.7 provides a comparison of the CD-ROM and MMCD systems.

10.4 Reference

1. CCITT, Recommendation H.261, "Video Codec for Audiovisual Services at $p\times64$ Kbits/sec," Geneva, August 1990.

Chapter

11

Computer Applications for Compression

11.1 Introduction

Within a PC, the audio and video may be compressed and expanded via software running on the central processing unit (CPU), by some dedicated hardware, or some combination of both. But before compression, the analog signals must be converted to digital data and routed to their destination. And after decompression, the digital audio and video streams must reach their respective destinations synchronously and without interruption.

So we will begin with a description of data conversion and information recovery, followed by a discussion of PC architectures for handling these signals. We will then look at software environments and algorithms for compression and decompression.

11.1.1 TV audio and video signals

In Chap. 2 we discussed the composite video signals in common use today. The source of these signals could be a demodulated signal from a TV antenna, a cable TV set-top box, a videocassette recorder (VCR), or video camera. We showed how a conventional TV set recovered the black-and-white image, color image information, and timing signals required to display each horizontal line and each field of the picture. Video signals can also contain the stereo sound, closed-caption text, teletext, and other information, as illustrated in Fig. 11.1.

A computer will also need to recover much of this same information in order to store, process, and display the program material. While most TVs process the signals in the analog domain exclusively, most

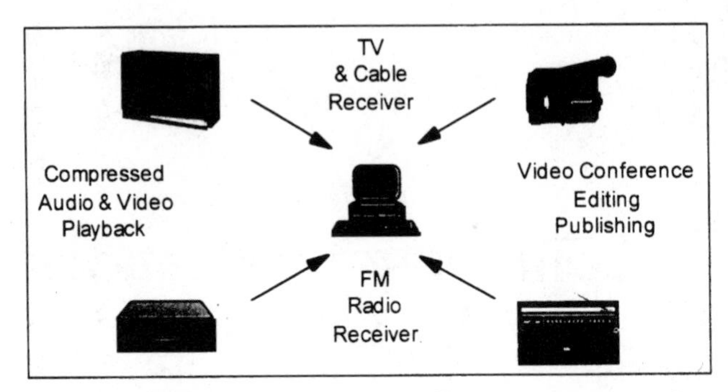

Figure 11.1 Multimedia architectures allow computers to receive, edit, and play back audio and video from a variety of sources.

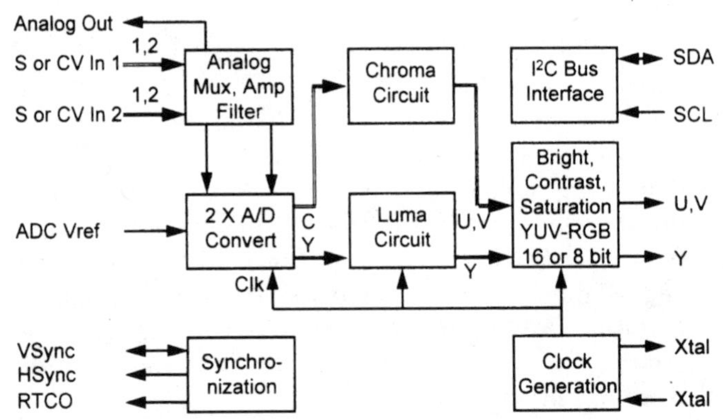

Figure 11.2 Block diagram of a digital NTSC and PAL analog composite or *S*-video decoder.

computers digitize the analog signal immediately and perform all subsequent processing digitally.

The composite video signal is filtered into black-and-white (luminance or *Y*) and color (chrominance) components. This format is often referred to as *S-video.* These signals are digitized with the chrominance information further separated into its orthogonal *U* (*B-Y*) and *V* (*R-Y*) color components. The computer must process this *YUV* image data or matrix into a decoded *RGB* signal, as shown in Fig. 11.2.

11.1.2 PC buses and video capture architectures

Audio and video capture, with its real-time high data rate (>20 Mbytes/s), requires the availability of sufficient bandwidth at the right time (Fig. 11.3). Live multimedia data streams of video and audio data are continuous and cannot be stopped. Data that are not

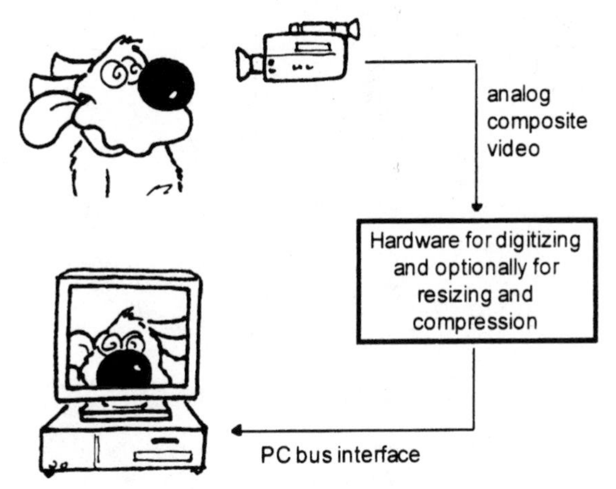

Figure 11.3 Video capture on a PC.

captured are lost, producing playback with distracting visual gaps and/or noise. Scheduling and allocating bandwidth in a predictable manner on the video bus is a major issue. In most consumer applications, audio has the highest priority because gaps in the data are most noticeable as unacceptable clicks. Video data are less sensitive to missing data, as previous images or portion of images can be repeated with little noticeable effect.

Several architectures exist to display live video signals on the PC. Currently, the PCI architecture with its bandwidth capability of 132 Mbytes/s is the only architecture allowing live video playback with the ability to capture and manipulate images on the PC. Older ISA- and EISA-based PCs had a prohibitive bus bandwidth of about 8 Mbytes/s. (See Table 11.1.)

11.1.3 Analog MUX architecture

An analog *RGB* MUX switches between the analog *RGB* graphics image and the live video image to display a live video window inside the graphics image. The video controller synchronizes itself to the VGA controller via timing signals transmitted through the *feature connector* of the graphics card. The video controller is also programmed to switch the *RGB* MUX at the desired line and pixel. This architecture is recommended when adding a video card to existing computers with basic graphic cards.

The drawbacks of this architecture include cost (additional of video controller, video frame buffer memory, DAC, and analog MUX) and the fact that video data never really enter the computer bus but are sent directly to the display (Fig. 11.4).

TABLE 11.1 PC Bus Architectures

Bus	Intro. date	Data width, bits	Clock rate, MHz	Peak throughput, Mbytes/s	Typical throughput*
PC	1981	8	8.33	8.33	0.5
ISA(PC AT)	1985	16	8.33	16.66	2
EISA	1988	32	8.33	33	8
NuBus	1987	32	10	40	20
	1993	32	20	80	50
DAV	1993	16	13–15	29.5	27
Microchannel	1987	32	10	40	8
architecture	1992	64	10	80	16
	1994	64	20	160	
VESA VLbus	1991	32	25–33	132	67
	1994	32	25–50	200	
	1994	64	25–50	4000	
VESA VMC	1994	8	25/33	25	23
	1994	16	25/33	50	47
	1994	32	25/33	100	95
VESA VAFC	1994	32	37.5	150	120
PCI	1993	32	33	132	65
	1993	64	33	164	130
	1995	64	66	164	130
Cardbus	1993	32	16	66	65
PCMCIA SFF	1994	32	33	132	65

*Not all measurements were available as this book went to press.

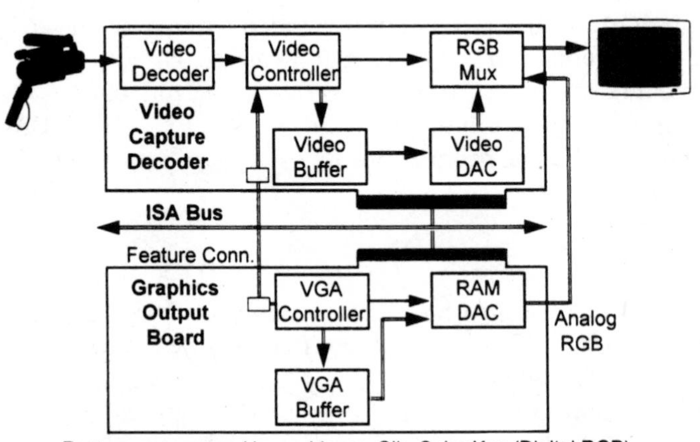

Figure 11.4 The analog *RGB* MUX switches between the analog *RGB* graphics and the live video in the window.

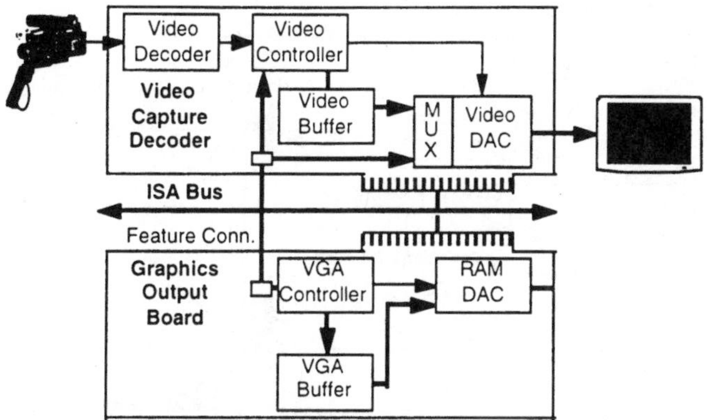

Figure 11.5 The video controller synchronizes the video windowing via the digital MUX.

11.1.4 Digital MUX architecture

Similar to the analog MUX architecture, a digital MUX is integrated into the video DAC on the video capture card. The digital *RGB* graphics image (sent via the feature connector) and the image from the video controller are combined to display a live video window. Timing signals are transmitted through the feature connector to allow the video controller to synchronize to the VGA controller, as illustrated in Fig. 11.5. This architecture requires intercepting the setting of the graphics RAM DAC and sending it to the video DAC. It has similar drawbacks to the analog MUX method.

11.1.5 DAC attach architecture

If the graphic card RAM DAC has two digital input ports and contains the VESA committee *advanced feature connector* (VAFC), a lower-cost video card (without video DAC and *RGB* MUX) can be built (Fig. 11.6). Synchronizing the graphics and video card is not easy, especially at high resolution and high pixel rates.

11.1.6 Shared frame buffer architecture

If the VGA graphics controller in the graphics card has a digital image port and if the card contains the VESA committee media channel (VMC) connector, an even lower-cost video card (without video DAC, *RGB* MUX, video buffer, and controller) can be built (Fig. 11.7). This architecture, like those preceding it, can only display video and are not able to edit, compress, or share the image with other devices in the computer or network. The availability of boards with VMC connectors is limited as of this writing.

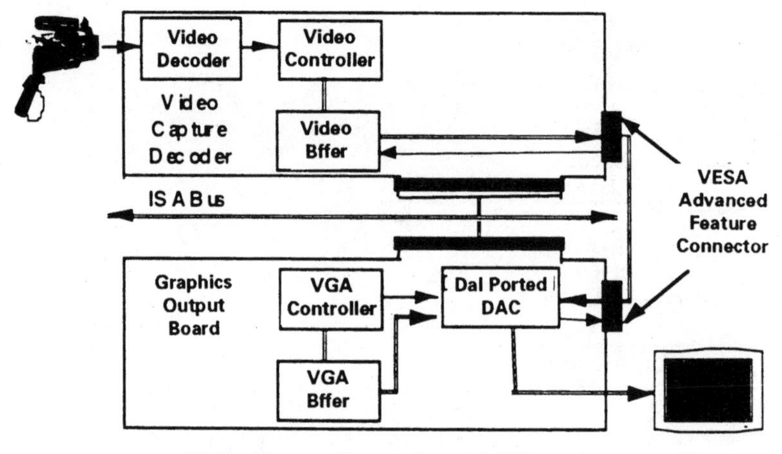

Figure 11.6 A dual-ported RAM DAC switches between the graphics and the live video.

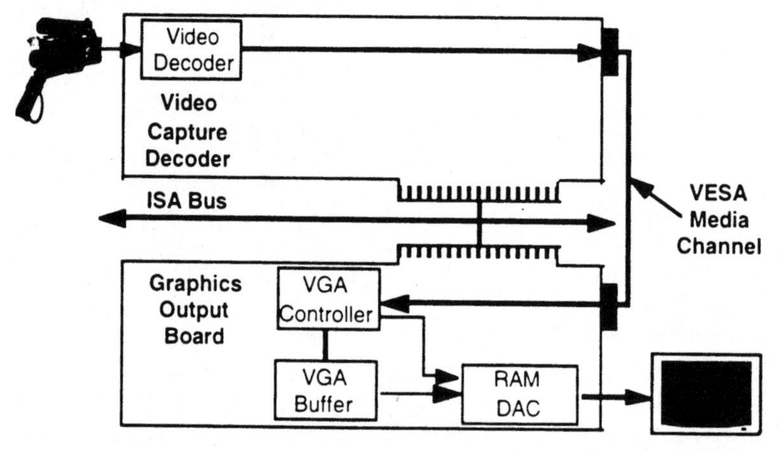

Figure 11.7 The live video image is written into the graphics frame buffer memory via the image port of the graphics controller.

11.1.7 Peripheral component interface (PCI) architecture

The PCI architecture frees the CPU to run the operating system on its local bus while video traffic is conducted on the peripheral PCI bus. The arbiter on the PCI bus establishes and manages the real-time importance of each master requesting the bus, increasing the robustness of the PCI bus and its capability to recover from errors and handle conflicting demands.

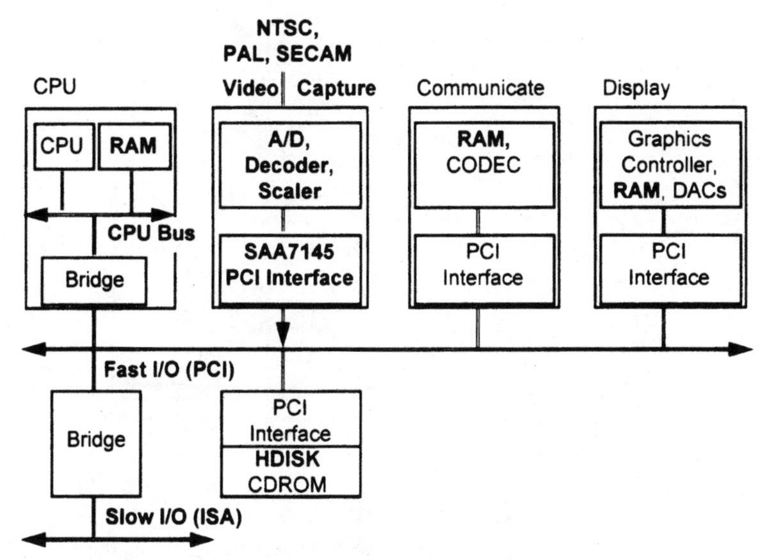

Figure 11.8 The odd and even fields of the interlaced video image are scaled and sent to various destinations in the PC for videoconferencing, editing, publishing, and playback applications.

The PCI architecture, with its bandwidth capability of 132 Mbytes/s, is the only architecture that allows live video playback and capture on the PC. The architecture uses a shared memory structure where the live images are decoded and scaled by the video capture engine and sent to the desired memory location in the computer. A block diagram of this architecture is shown in Fig. 11.8.

There are a number of benefits to this approach. For example, in a videoconferencing application, odd image fields from the local camera (the vanity image) can be sent directly to the frame buffer RAM of the graphics controller for display on the local monitor. Even fields are scaled down and sent to the videoconference compression engine RAM.

11.1.8 PCI multimedia interface

A PCI-based TV architecture can display programs as well as allow video capture for editing, desktop publishing, or videoconferencing. Television tuners accept the NTSC/PAL TV signal from a cable TV or antenna. The tuner selects one channel in the 50- to 850-MHz band and generates baseband composite audio and video signals. The video signal is digitized into the desired *YUV* square pixel or CCIR 601 formats. Simultaneously, the signal is sent to a closed-caption/teletext decoder to extract the associated information. The composite audio

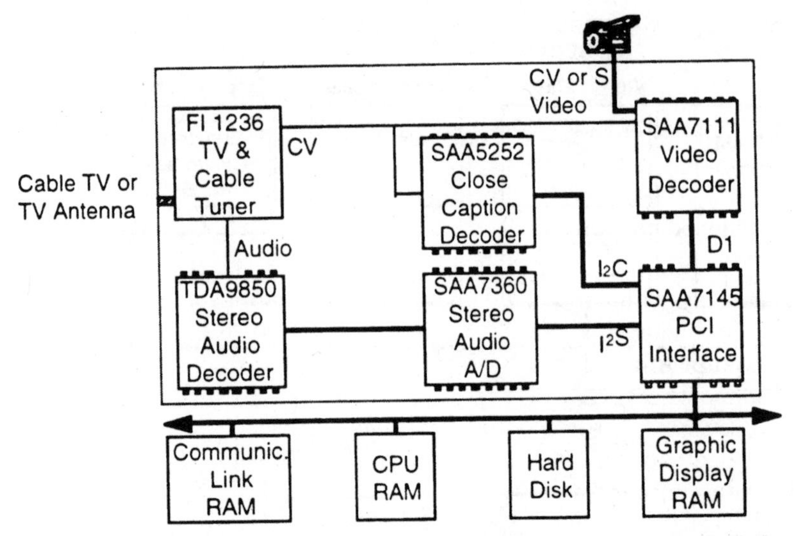

Figure 11.9 Video processing system: the TV image is decoded and scaled, the stereo sound is decoded and digitized, and the closed-caption information is decoded and sent to the PCI bus.

signal from the TV tuner is decoded into stereo sound and digitized by a stereo audio ADC.

A PCI interface IC scales the *YUV* image and transmits video, closed-caption information, and digital audio to the PCI bus. Because the component set requires no memory (beyond that already in the computer), it affords the smallest and lowest-cost video solution for "in the computer" solution (Fig. 11.9).

11.1.9 Stereo decoders

The multichannel sound (MTS) stereo signal from the tuner contains the left + right audio in the 50- to 15,000-Hz band. A pilot tone is transmitted at 15,734 Hz, and the left − right signal is modulated on a 31,468-Hz carrier. The second audio program (SAP) allows the TV viewer to listen to the program in one of two languages. The SAP sound is modulated on a 78,670-Hz carrier. The stereo decoder locks its PLL to the pilot tone and extracts the left, right, and SAP signals.

11.1.10 Video scalers

Scaling the video image is required for several reasons:

- The video image in the Windows environment on the computer monitor can be dragged and sized at the viewer's discretion.

- The image bandwidth must be reduced when writing to the hard disk.
- The size and format of the image need to be changed to match the requirements of the compression and graphic controller ICs.

11.1.11 Closed-caption and teletext decoders

Television program providers transmit images, sound, and text. The text contains closed-captioning for people with poor hearing and teletext containing business and other data. The text is sent as digital data during the vertical blanking interval (when no active video is transmitted).

The closed-caption and teletext ICs need to receive the analog composite video signal and locate the horizontal line where the digital data are embedded. The system then must recover the clock and extract the data sent. Next, error correction and detection takes place, and the text data are assembled and stored.

11.1.12 MPEG-1 audio and video decoder

The Philips SAA7131 MPEG-1 audio and video decoder performs system-level parsing, extracting the compressed video stream and the compressed audio stream and storing them in the external 2-Mbit DRAM first-in first-out (FIFO). The video and audio processors decode the compressed streams in the FIFO and assemble the decompressed frames and sound, as shown in Fig. 11.10. The video images are played out of the SAA7131 as *YUV* or *RGB* data in 352×288 (SIF format), 320×240 (CIF format), or 640×480 resolutions.

Using a modular PCI-based architecture, as described previously, new applications can be created to take these media one step further and allow complete access to video and audio data to edit, compress, transmit, and/or store.

11.1.13 Software decoding

Software decoding of compressed video and audio is promising because it is free; if you have already bought a personal computer with a powerful processor and lots of memory, you can see reasonable quality video in a window without buying any extra hardware (Fig. 11.11). There are two common environments for video software playback—*Quicktime* from Apple, and *Video for Windows* from Microsoft. For the purposes of example, we will examine Video for Windows.

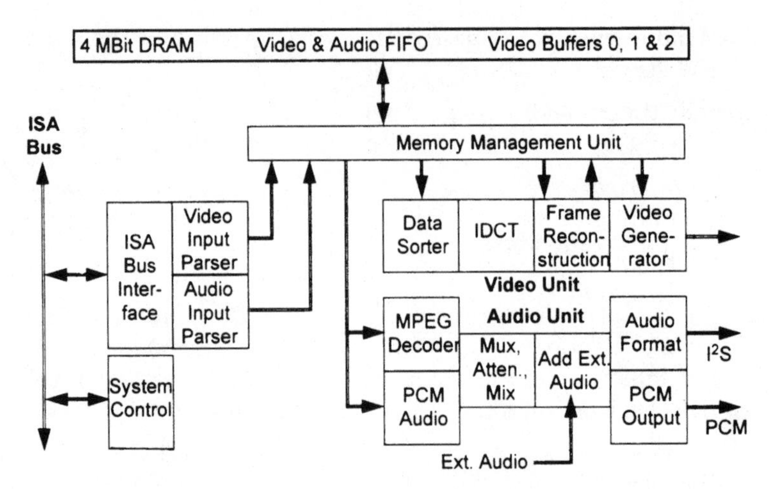

Figure 11.10 The MPEG-1 compressed audio and video data are sent via the ISA bus to the SAA7131 to be decoded into *RGB* or *YUV* digital video and I^2S digital audio.

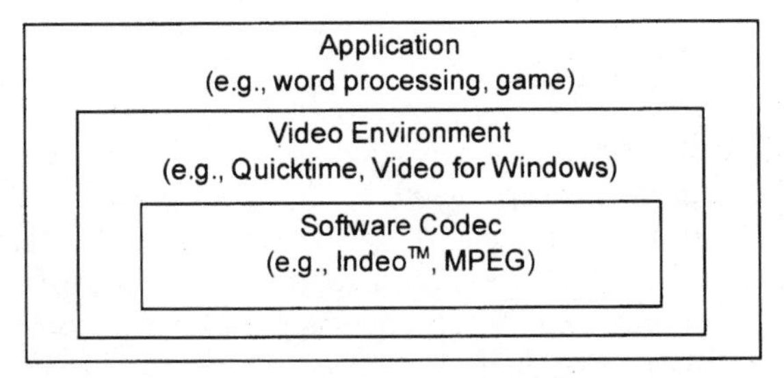

Figure 11.11 Model for software decompression interfaces.

11.1.13.1 Video for windows. Microsoft's Video for Windows (VFW) 1.0 was first released in November 1993, opening the door for inexpensive audio and full-motion digital video to millions of Windows PC users. Today, VFW provides a software mechanism for users to view, capture, play back, and edit video. VFW also provides developers a standard interface to create new digital video software applications, dynamically linked libraries (DLLs), and video device drivers. Some refer to VFW as a *video environment* for the PC. Microsoft's VFW contains playback applications and installable run-time DLLs (compression, decompression, file handling, media control, and other functions). Microsoft also distributes a software developers kit (SDK) for VFW. The following list contains some of the principle features of the Video for Windows SDK:

- Video for Windows sample applications and drivers that utilize and provide VFW services
- Definitions of the window classes for video capture, edit, and play back
- *Audio-video interleaved* (AVI) and *resource interchange file format* (RIFF) file-handling functions, interfaces, and file structures
- *Application program interfaces* (APIs) to enable the developer to write custom programs and routines
- Video CODECs, working with or without the *installable compression manager* (ICM)
- Audio CODECs, working with the *audio compression manager* (ACM)
- Video capture device drivers and/or MCI device drivers to perform step and streaming video capture and overlay functions

The flow of data and the modules used within VFW are slightly different based on the specific application and the function the user is trying to perform (view, capture, play back, or edit). We will start by explaining a simple capture application that allows the user to view and capture digital video data.

11.1.13.2 Video capture. A video capture application captures digital video data from a video source and either displays the video to the user and/or stores the data to disk. Possible video sources include a camera, television tuner, laser disk, and VCR (*S*-video or composite). The data can be captured a single frame at a time or as a real-time data stream. When allowing the user to view the data, the data can be viewed in either a preview mode or in an overlay mode. The *preview* mode of Video for Windows allows the periodic capture of a video frame and displays it. The rate at which the single frame is captured is dependent on the selected preview rate and the maximum throughput of the system being used. The *overlay* mode allows the capture hardware to write the video data directly to the VGA frame buffer via a feature connector or directly over the computer's PCI bus. Captured video data can be compressed and reformatted by a software codec prior to writing it to disk.

The following depicts primary data flows for VFW video capture applications. First the data are captured from the video source using the capture hardware and written either to the VGA frame buffer for overlay display or written to host memory for compression. The data can flow to be recorded to disk, clipboard, or decompressed for display (Fig. 11.12).

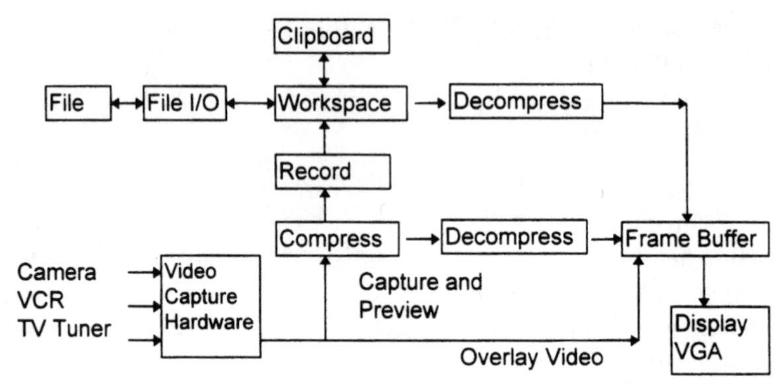

Figure 11.12 Video for Windows data flow.

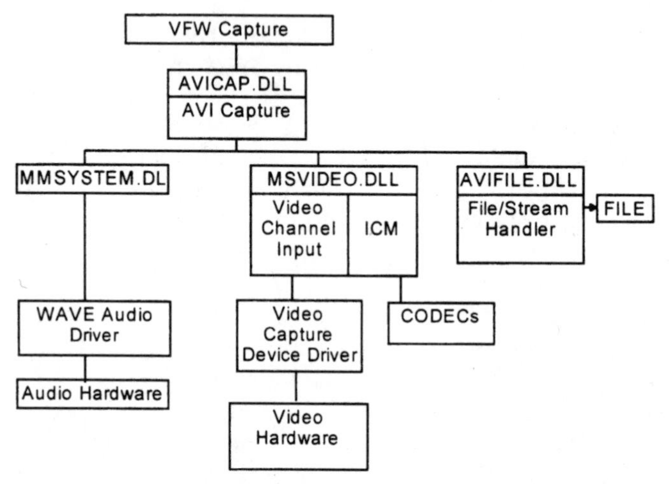

Figure 11.13 Video for Windows capture software architecture.

The Video for Windows application controls the flow of data and uses software DLLs and drivers created for Video for Windows applications to handle low-level capture, software codec compression and decompression, and data formats. This multimedia application is an integral part of a Video for Window's video capture device driver operation. The application is responsible for the allocation of memory used for capture and the management of the actual data buffers used for transfer. The application also is responsible for creating the AVI and AVI RIFF file and file headers.

The flow diagram of Fig. 11.13 depicts the basic structure of a VFW capture operation. Each block depicts an integral piece of the capture process. This software runs under Microsoft Windows, and uses

Windows messages to communicate between the application, DLLs, and drivers. Using the VFW software development kit, a developer can replace drivers and applications as desired, as long as they continue to utilize the provided APIs. This is possible, because each of the drivers use an installable-driver interface that processes the predefined low-level messages. The entry point for the installable-driver interface is the *DriverProc* function. For example, an application sends specific low-level video capture messages to the system, and the Windows system sends the appropriate message to the DriverProc routine of the video capture driver installed on the system. This standard interface allows developers to write their own custom applications and continue to use the provided DLLs, capture driver, and codecs. Developers may also choose to write their own software codec or custom video capture driver (supporting custom hardware) using the tools and APIs provided in the VFW SDK.

11.1.13.3 Video playback. Video playback is the process by which captured and stored digital video data are played back to the system screen. If the stored data are compressed, the data must be decompressed prior to display. A Microsoft VFW video playback application takes a generic VFW data file and attempts to render it on the system display. The data may be compressed in one of several possible formats. The player tries to see if the video device driver can do the decompression itself. For example, a JPEG compressed file is played on special JPEG decompression hardware. If the device driver is correctly implemented, the application will be responsible for very little of the work required to get the images to the screen. The application opens the device driver, issues some configuration commands, and begins providing the video driver with buffers of data for decompression and display. On the other hand, if the display driver cannot perform the decompression, the VFW player must determine what software components will be able to perform the decompression. The player may also need to determine if the image needs to be translated from one video-encoding format to another. In each case, the player tries to find a system component that will do as much of the work as possible. For example, if trying to play a VFW file on an 8-bit-per-pixel display, the player will try to find a decompression engine that can produce 8 bits per pixel as its output. As an alternative, the player could select a 16-bit-per-pixel format and translate the result using a dithering algorithm. The flow diagram of Fig. 11.14 shows the structure of the major VFW components and the flow of data and messages for a simple playback application where the video device driver is capable of decompressing the VFW data file via hardware, and the hardware is capable of sending the data to the VGA display.

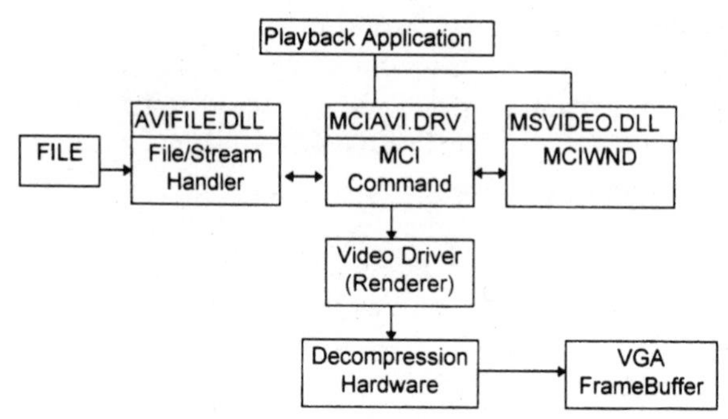

Figure 11.14 Video for Windows playback software architecture.

11.1.13.4 Video for Windows editing. Video editing is another application of Video for Windows. It also utilizes the MSVIDEO.DLL and the AVIFILE.DLL to produce modified video. The video is captured and stored as shown in the video capture section. After the video is stored, the data can be played back (decompressed) and modified. After editing, the data can be compressed and stored back to disk using the functionality of the AVIFILE.DLL. The SDK also contains APIs to specify and assist in video editing and special-effects work. Basic editing like cut, copy, and paste are supported by the AVIFILE.DLL.

11.2 Software-Only Compression Algorithms

Game and multimedia hardware makers and software developers alike are trying to sort through a crowded field of competing video compression systems for the winning algorithm for video playback on PCs and game platforms. The software-only compression techniques beginning to emerge in computer products are broadly based on three technologies: vector quantization (VQ), wavelets, and fractals.

11.2.1 Vector quantization

Vector quantization makes tables of representative values in an image during encoding and looks up those values in a decoding table during decompression. VQ codecs, such as Intel's Indeo 3.1, simply rearrange image data, selecting during decompression from among a smaller subset of allowable image features to represent the original. Though compute-intensive on the encoding side, VQ schemes can be decoded in software, without specialized hardware assists, for compression ratios of 5:1 to 100:1. The down side is that there is a tendency to "guesstimate" qualities in playback images. For example, in an image that contains 128 shades of green, a VQ-based codec might replace them all with a single average value in the middle of the color spectrum.

11.2.2 Wavelet transforms

Discrete wavelet transforms are essentially a variant on the discrete cosine transform technique of processing an image using a carefully selected set of basic functions. But where DCT-based implementations are limited to relatively simple cosine functions that require large amounts of multiplication and division, wavelets allow selection of a wider range of mathematical functions. The result is less computational complexity with no sacrifice in image quality. Compression ratios are about 15:1.

11.2.3 Fractals

While other methods operate on each individual pixel, fractal compression deals with full-screen images or large segments of them. The technique exploits the ability of so-called Mandelbrot equations to generate detailed, natural-looking images that break down into smaller domains that are in essence compressed versions of the same image; picture information can be viewed at any frame size, from thumbnail to full screen, with no observable loss in quality.

Fractals promise compression ratios as high as 1000:1. While still the province of a few experimental companies, such as Iterated Systems Inc. (Norcross, Ga.), the technology is receiving attention from major standards groups and is already being used in Microsoft's *Encarta* multimedia CD-ROM encyclopedia. Leading proprietary software-only compression algorithms are given in Table 11.2.

TABLE 11.2 Representative Producers of Software-Only Compression Systems

Algorithm	Company	Features
TrueMotion	Duck Corp.	Intraframe-only compression with features to reduce visibility of compression errors
Indeo	Intel	VQ, Intel will use new approach in the next release
Cinepak	SupeMAc Technology	VQ-based, used in few game applications
Wavelets	Aware Inc.	Hybrid of DCT and VQ
Fractals	Iterated Systems	Video editing applications

11.3 PC Applications

Personal computer designers have increasingly looked to integrate consumer product features into their computers. Current motherboard and add-on card manufacturers allow the consumer to watch TV or cable TV, listen to radio, and/or play their favorite audio and video CDs.

The PCI architecture goes one step further to allow computers to be used for videoconferences, edit still and motion images, and incorporate these images into interactive applications. Personal computer multimedia applications incorporating video are becoming more prevalent with the advent of more cost-effective video capture solutions and higher-performance processors. Several conflicting industry standards and proprietary video architectures are available, driving the end user to sort out the best cost/performance tradeoffs for their current and future needs. The ability to capture, manipulate (edit and/or compress), and display videos needs to be considered when selecting an architecture. Until the advent of PCI, a bus standard developed by Intel and others to address the requirements of multimedia and motion video interconnect, several hardware architecture standards competed to obtain market acceptance.

A family of PCI multimedia bridge products, starting with the SAA7145 and SAA7146, provide the capability to build video and audio capture systems that include interfaces for MPEG and videoconferencing hardware.

11.3.1 PCI bus

The PCI bus was discussed briefly earlier in this chapter, but it is appropriate at this point to expand on the technology because of its impact on video and audio compression systems in personal computers.

The standard for PCI bus bandwidth provides a distributed maximum transfer rate of 132 Mbytes/s. Possible devices vying for that bandwidth include storage devices, networks, graphics, and video subsystems. The ability of each device on the bus to receive and transmit bursts of information, on demand, greatly affects the capacity of the system as a whole. The symmetry of reading and writing between each device cannot be assumed.

In many first-generation PCI devices, developers interpreted the specification and provided products that, while adhering to the specification, did not take advantage of the PCI architecture's main benefit of high bandwidth by burst-mode transfers. Burst-mode transfer entails the collection of data at the source device until the PCI bus is available and only when a predefined amount of data are available;

TABLE 11.3 PCI Bus Efficiency Resulting from Burst Size and Latency

Arbitration latency (clock cycles)	1	3	3
Target latency (clock cycles)	2	2	3
PCI clock frequency, MHz	33	33	33
Burst size/transfer, bytes	PCI bus efficiency (maximum available bandwidth), % (Mbytes/s)		
1	25 (33)	17 (22)	14 (19)
2	40 (53)	29 (38)	25 (33)
4	57 (75)	44 (59)	40 (53)
8	73 (96)	62 (81)	57 (75)
16	84 (111)	76 (101)	73 (96)
32	91 (121)	87 (114)	84 (111)
64	96 (126)	93 (122)	91 (121)

the transmission of a single byte of data may not make sense when we consider the accompanying setup information. Because the data are transmitted from a buffer or FIFO memory, it is retimed to the maximum bus burst transfer rate of 132 Mbytes/s.

Recently, a host of ICs providing burst-mode capability have been announced. Steady-stream data sources, such as video, can now be repacked and retimed to take advantage of the higher bandwidth PCI bus, leaving bandwidth available for many other bus transfers. PCI targets are now capable of accepting these data rates. Burst-mode–capable graphic accelerators are available or have been announced from ATI, Chips and Technology, NeoMagic, S3, Trident, Western Digital, and Weitek, to name a few. These graphic accelerators can accept the video data with zero wait-state transfers and move it into the frame buffer for direct display. This ability is a key enabler for high-bandwidth transfers on the PCI bus.

Burst size, the number of bytes transferred in a single connection, can be shown to be the dominant factor in determining PCI bus utilization efficiency. Burst sizes of 32 bytes or more provide utilization efficiency of over 90 percent, while single-byte bursts cause the bus to work at less than 25 percent efficiency. This easily explains why with first-generation PCI video implementations, with many graphic controllers capable of accepting only 1 double word (*Dword* or 4 bytes) per burst transferred, it appeared as if video was taking up a large percentage of the bus and degrading the entire system performance.

As expected, when the data per PCI transfer become small, the transfer setup latencies become more influential on bus efficiency. As shown in Table 11.3, second-order effects include:

- *Arbitration latency.* The number of clock cycles to establish a source for data transfer

- *Target latency.* The number of clock cycles to establish a destination
- The number of block transfers for a given connection

Data streams that are transmitted on the PCI bus can now be examined as a percentage of the typical available bandwidth:

- *Video.* Full-scale, full frame rate interlaced NTSC video (640×480 resolution, 16 bits/pixel, 30 frames/s) = 18.4 Mbytes/s or 15 percent. Scaled, CIF MPEG-1 (320×240 resolution, 16 bits/pixel, 30 frames/s) = 4.6 Mbytes/s or 3.8 percent.
- *Audio.* Stereo, CD-quality (48 ksamples/s, 16-bit resolution) = 192 kbytes/s or 0.16 percent.
- *Networking.* SCSI = 10 Mbytes/s or 7.6 percent. Ethernet = 10 Mbytes/s or 1.0 percent.

It is easy to see that all these devices can be simultaneously supported on the PCI bus, with plenty of bandwidth available to the CPU.

11.3.2 PCI video enables new applications

As previously stated, there are many alternatives for handling video in a PC. The appropriate architecture should be determined by versatility, performance, and cost. Videoconferencing, compressed video storage and playback, and home video editing are just a few applications that require video to have direct access to the PCI bus. The ability to perform closed-caption, teletext, or other vertical blanking interval decoding in software necessitates direct access to PCI as well.

Several graphic accelerators provide video ports to establish quick and inexpensive video solutions. However, in order to send video data to the system for manipulation and/or storage, the data must first be sent to the graphics frame buffer and the system must fetch the data from the frame buffer. Nonsymmetrical read and write efficiency for individual PCI devices has a tremendous effect on the overall system performance. Read performance is often far worse than write performance. A 20 percent efficiency (1 Dword of data transferred an average of every 5 clock cycles) is experienced due to arbitration and target latencies and the inability of the graphics controller to provide burst data from the frame buffer to the PCI bus. This causes the PCI bus utilization for video to soar to almost 70 percent (effective 92 Mbytes/s equivalent bandwidth to transfer 18.4 Mbytes/s, full resolution, full frame rate NTSC data). This video latency is particularly cumbersome for real-time compression applications such as videoconferencing. The effect is fewer video frames per second or a video image the size of a postage stamp.

As far as cost is concerned, PCI video works on a shared frame buffer architecture and requires no local memory. It can burst the video information to any memory address in the system as well as the graphics frame buffer. A typical PCI video-audio capture board has been shown to take up less than 8 in^2 of board space and consists of only three IC devices—video capture demodulator, audio ADC, and PCI bridge chips.

11.3.3 The SAA714X family of PCI multimedia bridge ICs

To provide flexibility to system designers in the applications they support, a versatile PCI multimedia bridge architecture has been developed. The SAA7145 and SAA7146, the first two members of this family, are full-bandwidth PCI bus master devices, capable of symmetrical, bidirectional transfers in burst mode. Data can be saved in one of the various on-board FIFOs until it can be processed elsewhere in the system. In addition to handling video data streams, both devices also support audio capture and parallel slave devices. This architecture can individually direct the various data streams to any memory target within the PC system.

The SAA7145 and SAA7146 interface directly to any D1 style (CCIR-656) or *YUV*16 video source, such as single-chip video decoders like the SAA7110 or SAA7111, digital camera outputs, as well as decompress video provided by MPEG or videoconferencing engines. The digital data can be transferred directly to the PCI bus or can be color space transformed to meet the required format of the target graphic device. On-board scaling and filtering to any arbitrary size allows dynamic window (drag and drop) sizing. Down scaling can also be used as a means of further reducing bus bandwidth utilization. Mask clipping, the ability to occlude the video with arbitrarily shaped windows, allows the destination device to dictate what parts of the video information is displayed. Instead of stopping or slowing video every time another window partially occludes an active video window, mask clipping allows the video to continue in the remaining display region without any noticeable degradation.

Both devices also have the capability of capturing audio, whether through an audio ADC connected to a live video source or from a decompressed MPEG or videoconferencing stream, through an I2S serial, digital audio interface. The audio signal has a separate DMA channel for independent transmission to any destination for editing or output.

For typical CD quality audio of 44.8 ksamples/s with 16-bit resolution for each channel, the PCI bus utilization for the audio is less that 0.2 percent.

To support specific applications whose hardware currently sits on slower 16-bit parallel buses, a bidirectional data expansion bus interface (DEBI) has been developed that can be configured by software to look like either ISA or Motorola-style interfaces. MPEG-1 and MPEG-2 codecs, as well as H.320 and H.324 videoconferencing codecs can be integrated into the audio-video subsystem. By doing so, users will have the same flexibility of outputting, manipulating, and storing these data sources as live sources. At maximum ISA transmission rates of 8 Mbytes/s, PCI bus utilization is only 6.1 percent.

The SAA714X incorporates an intelligent-device architecture that allows on-the-fly reprogramming. Each multimedia bridge device can reprogram itself without the intervention of the CPU to accomplish tasks that would be prohibitive due to interrupt and software latencies. The register programming sequencer (RPS) is able to reprogram itself on up to a line-by-line basis. This feature is important to deal with VBI closed-caption data separate from other video information. Other applications include separating every other field for display and compression (for videoconferencing). Each destination can maintain separate and distinct scaling, format, clipping, and other attributes.

The performance and versatility of the new generation of PCI video clearly breaks the perception molds that many have set for this architecture. PCI bandwidth is available to achieve full resolution, full frame rate video, without interfering with other system activities. In addition to displaying video, new multimedia applications can be enabled to manipulate, compress, and store video images that competing architectures cannot. While performance is provided, this cost-competitive solution promises to be a major building block for architectures to come.

Index

Index

Index note: The *f.* after a page number refers to a figure; the *t.* to a table.

ABOUT THE AUTHOR

Stephen J. Solari is at C. Cube Microsystems in Milpitas, California.